ROBERT FLUDD AND THE END OF THE RENAISSANCE

In lumine tuo Videbimus lumen.

Si tu illustres lucernam meam, Iehova Deus Splendentes efficies tenebras meas.
Pf. 18. 29

Portrait of Robert Fludd

ROBERT FLUDD AND THE END OF THE RENAISSANCE

WILLIAM H. HUFFMAN

ROUTLEDGE

London and New York

First published in 1988 by
Routledge
11 New Fetter Lane, London EC4P 4EE
29 West 35th Street, New York NY 10001

© William H. Huffman 1988

Typeset in 10 on 12 point Baskerville by
Columns of Reading
Printed by Richard Clay Ltd.,
Bungay, Suffolk

British Library Cataloguing in Publication Data
Huffman, William H.
Robert Fludd and the end of the
Renaissance.
1. General knowledge. Fludd, Robert,
1574–1637. Biographies
I. Title
082'.092'4

Library of Congress Cataloguing in Publication Data
Huffman, William H.
Robert Fludd and the end of the Renaissance/William H. Huffman
p.   cm.
Bibliography: p.
Includes index.
1. Fludd, Robert, 1574–1637. 2. Renaissance—England.
3. England—Intellectual life—17th century.
4. Occultism—England—History—17th century.
5. Hermetism—History—17th century.
I. Title.
BF1598.F58H83 1988
133'.092'4—dc19   88–717

ISBN 0–415–00129–3

To William John, with love

CONTENTS

ILLUSTRATIONS

ACKNOWLEDGMENTS

I would like to gratefully acknowledge the contribution of some of the many who made the completion of this book possible.

Sarah Huffman's diligent efforts greatly aided the final preparation of the first draft of the manuscript, and Margaret V. Van Dyke's help with the second is most appreciated. Professor Charles G. Nauert, Jr provided continued encouragement from the inception of the project and valuable criticism. Research support was underwritten in part by a Frank and Louise Stephens History Fellowship Award and an Allen Cook White, Jr Fellowship Award through the History Department at the University of Missouri-Columbia. The support of the department itself in other ways was of great value to me.

I would like to sincerely thank Ann Edwards of the University of Missouri-Columbia Library for her untiring and dedicated help in acquiring needed books and other materials; Anne Oakley of the Cathedral Archives and Library, Canterbury, who uncovered valuable information on the Fludd family; Dr Felix Hull, Archivist, Kent Archives Office; and Mr D. A. Rees, Archivist, Jesus College, Oxford. I am also grateful for the courtesy and helpfulness shown me by the staffs of the British Library, Bodleian Library, Trinity College, Cambridge, Library, Wellcome Institute Library, Public Record Office, Warburg Institute, and the Library of Congress.

Joscelyn Godwin kindly permitted me to use a number of plates from his own book on Fludd, for which I am most grateful. I wish to thank the Bodleian Library for permission to reproduce Fludd's signature, and the Wellcome Institute for the History of Medicine Library permission to reproduce photographs of works in their collection. Permission was graciously given for the reprint of the 'Declaratio Brevis' translation through the courtesy of *Ambix*, where it first appeared. Finally, I would like to thank Robert Greene, Rector of Otham, and his wife Diane, for the warm hospitality they extended during a research visit to the parish church.

xi

ABBREVIATIONS

APC	Great Britain. Privy Council. *Acts of the Privy Council of England*
CPR	Great Britain. Public Record Office. *Calendar of Patent Rolls*
CSPD	Great Britain. Public Record Office. *Calendar of State Papers, Domestic Series*
CSPF	Great Britain. Public Record Office. *Calendar of State Papers, Foreign Series*
DNB	*Dictionary of National Biography*
HMC	Great Britain. Historical Manuscripts Commission. *Reports*
HMSO	Her Majesty's Stationery Office
JWCI	*Journal of the Warburg and Courtauld Institutes*
LI	Great Britain. Public Record Office. *Lists and Indexes*
DHS	Oxford Historical Society Publications

INTRODUCTION

The subject of this book, Robert Fludd (1574–1673), provides us with an excellent opportunity both to gain insights into the world of the late Renaissance, and to better see, and perhaps understand, the larger perspective of the great revolution in thought which took place at that time, and its consequences for our own. Fludd was integrally a part of a cultural tradition which died with him: that thread of the Renaissance stemming from the Florentine Academy, which encompassed the Latin translations of Plato and Hellenistic Neoplatonist works by Marsilio Ficino, and from Pico della Mirandola's attempt at a grand synthesis of all philosophy and religion. There is a real sense in which it is justified to say that the Renaissance as a whole, considered as an intellectual movement seeking to unify and reform society based on the wisdom of classical Greece and Rome, was rapidly passing from the scene at the end of Fludd's life. No one would deny that radical changes took place in the course of the seventeenth century, and it is precisely because Fludd made his great exposition of Renaissance Neoplatonist science and philosophy, which was truly a summation and synthesis of centuries of traditional thought, at the critical point of change in this period that his work may be seen as the conclusion of an era.

The reaction to his books mostly divided into favorable judgment from those who adhered to the old traditions, and rejection from those who were turning away from them to discoveries based on empirical, mathematical models instead of ancient religious and metaphysical ones. The other demonstrable fact showing that Fludd stood at the end of his tradition is the rapidity of the decline of interest in his books, with a corresponding rise in uncomplimentary opinion, much of which has continued unchanged to the present.

This theme is explored by developing the background context of the Late Renaissance, and showing that his ideas, which were derived from his early education and shaped at Elizabethan Oxford and by six years' travel in Continental Europe, were a product of his time; that he was a well-respected, prominent and orthodox

1

physician in Early Stuart England during and after the time his books were published; that his circle of friends and acquaintances included some of the most important people of his time; that he was not a member of the Rosicrucians, as often alleged; that acceptance of his ideas was underscored by his royal patronage; that crucial changes were taking place in a mixture of new and old in his debates with Kepler, Mersenne and others; and that the fabric of his philosophy, which was completely woven from Renaissance Christian Neoplatonism, was necessarily doomed when science and philosophy were divorced from religion and metaphysics derived from ancient sources.

Robert Fludd was a diminutive, unusually studious son of a wealthy and prominent knight, and was born and bred in the world of Late Elizabethan England. At his country manor in Kent, probably from his own experiences, and with the encouragement of a tutor, he developed an interest in the divinatory arts, music and other disciplines, and was also possibly introduced to the idea of their Pythagorean-Platonic interrelatedness and an organizing scheme encompassing them. It was also the time and place where he acquired his intense religious piety. Thereby we see a pattern which never changed: that of a scholarly and pious nobleman who was deeply committed to the religious, philosophical and scientific truth of a thoroughly Renaissance cosmic view, which followed in a long tradition of Gnostic, Pythagorean, Platonic and allied systems of thought. His devotion to the unquestioned truth and divinity of this tradition set the stage for his studies, lengthy publications, associates, controversies and place in history.

In his own liftime, Fludd was continuously engaged in disputes over the structure of his philosophy, and research into those with whom he was engaged in debate, and the grounds of the debates, has provided fruitful information for studies of Fludd by others, and certainly lends weight to the thesis presented here. By the time of Fludd's systematic and elaborate summary, complete with many carefully executed, fascinating diagrams and illustrations, the tradition in which he worked was beginning to be rapidly undermined and was soon engulfed by a radical change in the intellectual climate. As with all change in history, this one did not take place overnight nor succeed in eliminating completely all the adherents to this old tradition; but there is no question that the seventeenth century witnessed much of the transition from the ancient world-view to the modern one.

Robert Fludd's unique and important accomplishment was to produce in his works a grand summation of Renaissance Christian Neoplatonist thought, which encompassed two millennia of ancient, medieval and Renaissance traditions in the arts, sciences and medicine in a religious and philosophical context. It is also significant that he produced these works not as an outsider or wandering outcast, as did, for example, Paracelsus or, to an extent, John Dee. Though much of Fludd's metaphysics and descriptions of the divinatory arts are today considered occultist (which for him was simply the way God's spirit emanated through and worked in the upper and lower realms), he was an insider, an established member of conventional society: a man of noble descent, wealthy son of a knight, Oxford-educated, member of the College of Physicians of London and successful physician, with English kings as his patrons, sometime tutor to and friend of high nobility and clergy as well as contemporary physicians, patentee to make steel, from his own invention, in England. His works were clearly the greatest expression and summation of Renaissance Christian Neoplatonist thought, which many of his contemporaries admired. But just as a star burns brightest nearest the end of its life, Robert Fludd's great intellectual and artistic achievement was completed at the end of the era of which he was so much a part and which he expressed so well in his unparalleled volumes.

Chapter I

THE EARLY YEARS: ELIZABETHAN ENGLAND AND CONTINENTAL EUROPE

> How can a real philosopher enucleate the mysteries of the Creator in the creature, or judiciously behold or express the creature in the Creator but by such rules or directions as the onely storehouse of Wisdom, namely the holy Scriptures have registered . . .
>
> Fludd, *Mosaicall Philosophy*

One of the main problems confronting someone interested in Robert Fludd is the lack of information about his formative years, as well as about his later associations. While there is much remaining that we do not know, a combination of new materials and old can be put together to expand what we do know, so that a better picture of this pivotal figure of the age may emerge.

Robert Fludd was born at his father's country manor, Millgate House, in the parish of Bearsted, Kent, sometime shortly before 17 January 1573/4, when he was baptized in the parish church.[1] He was descended on his father's side from gentry who were originally Welsh (they went by the related name, Lloyd), but who lived in Shropshire for at least three generations before Sir Thomas Fludd, Robert's father, acquired Millgate House.[2] Throughout his life, Robert remained intensely proud of his family heritage, making sure that the title of Armiger or Esquire appeared directly after his name and before the title of Doctor of Physic in all his printed works.

In his attack on Fludd for advocating use of the "weapon-salve", (discussed more in Chapter V), Parson Foster complains about this order of titles:

> . . . being a weapon-bearing Doctor, may well teach the weapon-curing medicine, especially setting the Armiger before the Doctor, the Gunne before the Gowne, the Pike before the Pen.[3]

Fludd smarted from this remark, and replied:

4

[Good Reader] observe this unreasonable jest . . . verily I
think . . . it is certaine that now adayes a reasonable Esquire
thinketh much to yeeld place unto a Doctor: But this is not our
question . . . I put the Esquire before the Doctor . . . for two
considerations: first, because I was an Esquire, and gave armes
before I was a Doctor, as being a Knight's sonne: next because,
though a doctor addeth gentilitie to the person, who by descent
is ignoble; yet it is the opinion of most men, and especially of
Heraulds, that a Gentleman of Antiquietie, is to be preferred
before any one of the first Head or Degree: And verily for mine
own part I had rather bee without any degree in Universitie,
than lose the honour was left me by my Ancestors.[4]

The value Fludd placed upon his ancestral heritage is graphically
obvious in the only two portraits of himself that he commissioned,
appearing as the frontispieces of the *Philosophia Sacra* of 1626 and
the *Integrum Morborum* of 1631. In the first, he is flanked by a large
coat of arms which displays no less than twelve quarterings; it is
one of the most prominent features of the portrait (frontispiece).
The second has a similar arrangement, this time showing an older
man and displaying only six quarterings (Plate 20). A copy of the
second portrait appears in Boissard's *Bibliotheca* of 1630, along with
high praise for Fludd's descent from so noble a family; the article
was probably written by Fludd's publisher at the time, William
Fitzer.[5]

The Fludd family rose to prominence in Kent and at Elizabeth's
court through the efforts of Robert's father, Sir Thomas Fludd. Sir
Thomas was trained in the law,[6] and entered the crown service as
a young man, where he served as victualler for the army in the
garrison towns of Berwick in England and Newhaven, France, in
the early 1560s.[7] It must have been shortly after this service that
the elder Fludd acquired the manor of Millgate and married
Robert's mother. Sir Thomas was appointed Justice of the Peace
for the Bearsted district in that same decade, an office he held until
his death in 1607.[8] In 1568, he was given a grant for life as
Surveyor of crown lands in the County of Kent and the cities of
Rochester and Canterbury,[9] and in the mid-1580s he was
appointed Receiver of crown revenues for Kent, Surrey and
Sussex.[10] In 1589, the senior Fludd was made the paymaster for
the four thousand troops sent by Elizabeth to aid the French

King.[11] His performance in this post brought him a knighthood during January, 1589/90, the month the English force returned home.[12] Back in England, the newly elevated knight continued to carry out numerous special commissions for the crown,[13] and a few years later the now elderly Fludd was called back into service as Paymaster by the Queen. It seems that the long-time Paymaster to Her Majesty's Forces in the Low Countries, Sir Thomas Sherley, defaulted on payments to the troops and declared bankruptcy. Sherley was put in jail, and in 1597 a very angry Queen personally summoned for the post a man she knew to be of high integrity and honesty: Sir Thomas Fludd.[14] Court intrigue, the ambition of younger men to make a name at court, and Fludd's insistence on having a company of troops at his command to guard the treasure all worked against him, however, and the records show that he held the office for only one month, April-May, 1597.[15] Sir Thomas continued to maintain favor at court, however, and his stewardship of some twenty manors in Kent was renewed by letters patent when James I ascended the throne in 1603.[16]

The Kentish Fludd patriarch also sat in three Parliaments, in 1583, 1597–8 and 1601. Kent returned ten members, two from the county, two each from the cities of Rochester and Canterbury, and two each from the boroughs of Queensborough and Maidstone; Sir Thomas was a member for the latter.[17] He was also afforded the honor of admission to Gray's Inn on 9 March 1600/01, a good indicator of the status he had attained.[18] Thus Robert's father had firmly established the family's wealth and prestige in Kent by the time the youth began his studies at Oxford in 1592. The family seat had been enlarged and improved, which transformed it into a grand manor befitting the position of its owner.

Robert's family was a large one. Sir Thomas married Elizabeth, the daughter of Philip Andrews of Wellington in Somerset, Esquire, about 1565. In all, the Fludds had twelve children, nine of whom lived to adulthood: six sons, Edward, Thomas, William, John, Robert and Philip; and three daughters, Joan, Katherine and Sarah.[19] By the time of Sir Thomas' death in 1607, only the three daughters and Thomas, John and Robert were still living.[20] Elizabeth Fludd, Robert's mother, died in 1592, the year he entered Oxford;[21] Sir Thomas was remarried to Barbara, daughter of Matthew Bradbury of Essex, Esquire, the widow of Sir Henry Cutts.[22] The only two surviving members of Robert's immediate

6

family when he died in 1637 were his married sisters Katherine and Sarah.[23] (See genealogical table, Appendix D.)

Unlike his older brothers, Robert decided not to study the law. His early education was probably received from tutors at the family manor.[24] From there he followed his brother, John, two years his senior, in matriculating at St John's College, Oxford, as a commoner, age 18, on 10 November 1592.[25]

Oxford had undergone considerable change during the course of the sixteenth century, yet it had retained much of its medieval, pre-Reformation and pre-humanist structure. In the early part of the century, the humanists were unanimous in decrying the sterility of the medieval scholastic methods and the content of the curriculum.[26] The English universities were ecclesiastical institutions, under both the King and the Pope, with the latter superior in authority. Their primary task was to prepare churchmen for office.[27] Under this arrangement, it was unseemly for a young man of gentle birth to attend the university unless he was one of those few who intended to seek church office.[28]

Before the middle of the century, however, some profound changes took place at both Cambridge and Oxford. Humanist ideas about needed changes in the curriculum began to take hold. Within the tradition of the seven liberal arts dating from Roman times, emphasis was shifted from medieval exercises in logic to the study of advanced Latin grammar, Greek, rhetoric and moral philosophy, the better to emulate the great Greeks and Romans. Between 1500 and 1520, the Bishops Fisher and Foxe introduced lectures in all four of these humanistic areas at both universities, and in the years 1511–13 Erasmus himself lectured at Cambridge.[29] Henry VIII endowed professorships in Greek and Hebrew at the universities, and under Edward VI the entire liberal arts curriculum was revised to bring it in line with humanist ideals.[30] These changes remained intact under both Mary and Elizabeth.[31]

Not only did the curriculum undergo changes, so did the composition of the student body. With the changing bureaucratic needs of government, the lack of traditional military opportunities, and the changing fashions of a Renaissance society extolling the virtues of the learned noble, the universities were able to lure well-born sons to their halls in greater numbers. Oxford and Cambridge had become significant centers of humanist education, and increasingly succeeded in convincing the upper classes that such a course of study for their sons "not only instilled virtue and wisdom but nourished the capacities and talents to manage the affairs of state."[32] In Elizabethan times the number of students of noble or gentle birth climbed dramatically, until they reached a slight majority by 1600.[33]

Another profoundly significant event affecting the universities was Henry VIII's break with Rome and the subsequent establishment of Elizabeth's unique form of Reformation. The paradoxical consequence was that even though the universities legally became lay corporations under the direct control of the crown, usually had lay Chancellors and provided new opportunities for those who did not intend to seek careers in the church, they still remained essentially ecclesiastical institutions. Under the Tudors the statutes remained in their medieval form, and were even followed by the founders of new colleges in the sixteenth century.[34] The majority of crown commissioners visiting the universities continued to be churchmen in high office, the colleges were headed by men in orders, and the majority of the Fellows were still required to be in orders or take them at the end of a designated period of time after election as a Fellow or after receiving the M.A.[35]

St John's College, Oxford, where Fludd matriculated in 1592, had followed this pattern. The College was founded by Sir Thomas White, a layman, former Master of the Merchant Taylors' Company in London, and, when he could no longer avoid the burden of public office, an Alderman and Lord Mayor of that city.[36] In May, 1555, he received letters patent from Philip and Mary for the founding of a college to be named after the patron saint of the Merchant Taylors, thus St John Baptist College. This was the time of the reimposition of Catholicism in England, and Sir Thomas White, himself a Catholic, founded the College with the express purpose of improving the education of the clergy and hence contributing to church reform.[37] The statutes were mostly patterned after those of Corpus Christi, another recently founded college. Consistent with the now-prevalent humanist educational ideals, the founder endowed three lectureships: Greek, rhetoric (serving as advanced Latin) and logic or dialectic.[38] In 1580, a lectureship in civil law was added (by this time degrees in canon law had been dropped), and two years later one in natural philosophy.[39]

After 1583, the College consisted of the President and fifty

Fellows, thirty-seven elected from the Merchant Taylors' School in London, two each from Bristol, Coventry and Reading schools, one from Tonbridge, and six reserved for the founder's kin.[40] Between 1578 and 1600, which includes the period when Fludd was there, about twenty commoners were in residence.[41] According to the founder's wishes, one-fourth of the Fellows were to study the law, only one should study medicine, and the rest were to study liberal arts as a preparation for theology.[42]

Upon arrival, usually at the age of sixteen to eighteen, the new student had his name entered in the Buttery-book and the Matriculation Register, signed the Subscription Book, appeared before the Vice-Chancellor and took oaths accepting the Thirty-Nine Articles of Faith, the Prayer Book and the Royal Supremacy.[43] The important tasks of deciding a course of study, assignment to a Faculty and choice of a tutor were performed. The tutor not only supervised the studies of his charge, he also was supposed to look after his student's finances, morals and manners. In this capacity he did not act as a college official, but was normally paid by the student's parent, guardian or fellowship.[44] Fludd started his studies in the Arts Faculty, and, as he tells us later, his tutor was John Perrin (1558–1615), a Doctor of Divinity and the lecturer in Greek at the College.[45]

The undergraduates and bachelors were required to attend the appropriate college or public lectures unless excused by ability or progress. At St John's, the logic lecture was delivered immediately after mass, about 6.00 a.m., Greek at 9.00 a.m., and rhetoric at 1.00 or 2.00 p.m., with the participants permitted to question the lecturers on difficult points.[46] All scholars were also required to take part in disputations appropriate to their standing and course of study. Although commoners did not have the threat of loss of their fellowship for lack of attendance at lectures, participation in disputations and maintaining personal discipline, they could be fined or disqualified from taking their degree if they did not comply with the rules.[47]

In fact, most commoners did not take degrees, probably because many of the young nobility only came to the university because gaining exposure to university life was the fashionable thing to do, not because they felt the necessity to obtain a degree in order to be a country squire, merchant or adventurer. In this respect, Fludd stands out as unusual. As commoners increased in social rank, fewer took degrees. For example, in the fifty-year period 1598–1648, less than half of all the commoners at St John's proceeded to a degree there or elsewhere.[48] During this span of time 51 have been identified as sons of peers or knights (as Fludd was), and only 7 of this class took degrees. In the same period only 19 of 72 sons of esquires finished a degree. By contrast, half of the sons of gentlemen, 23 of 32 sons of clergy and 134 of 195 plebeian sons completed a bachelor's degree.[49] By these figures it is obvious that Fludd was exceptional for a knight's son in the interest he showed in his studies.

Fludd is even more unusual in having studied medicine at Oxford. In the thirty-year span from 1571 to 1600, less than 50 medical degrees of all kinds were granted from Oxford, and only 35 licenses to practice were issued, just over one per year.[50] Of the 33 Fellows of the College of Physicians in London in 1618, it is known that 27 took arts degrees from Oxford or Cambridge, but only 13 went on to the medical degree in England, and only 5 of these were from Oxford.[51]

It is my belief that Fludd's interest in medicine, his characteristic science and philosophy (founded on theology and Renaissance occultism) and the work of gathering and classifying the sum of man's exoteric and esoteric knowledge all derive from his days at St John's.

Fludd's interest in medicine quite likely stems from the fact that St John's College was one of the very few colleges in England which had any provision for a medical fellowship, and thus normally had a Medical Fellow in residence. This influence can be seen by the fact that of the five Oxford M.D.s in the Royal College of Physicians in 1618, three were graduates of St John's; another member, Sir William Paddy, James I's Physician-in-Ordinary, had taken his B.A. from this school.[52] The Medical Fellow during Fludd's stay at St John's was Matthew Gwynn (1558–1627), one-time reader of the music lectures, poet, playwright, orator and later Professor of Physic at Gresham College.[53] Gwynn was also to be a colleague of Fludd's in the College of Physicians in London. In 1611, he wrote a tract against a London practitioner which showed that, although he was primarily a Galenist, he was also quite familiar with the main Paracelsian medical works.[54]

Another very likely influence on Fludd's life at St John's was Ralf Hutchinson, President of the College from 1590 until his death in 1606. Hutchinson had been Matthew Gwynn's predecessor as Medical Fellow from 1581 to 1586, and from 1579 to 1581 had been the rhetoric Reader.[55] In Hutchinson the mixture of a classical education, Anglicanism and medicine is clear, and, since he was President during the entire term of Fludd's studies at St John's, he may well have taken a personal interest in the unusually studious commoner knight's son. Though he had studied medicine for five years, Hutchinson apparently forfeited his fellowship to marry, then took holy orders, held three vicarages and eventually took a B.D. in 1596 and a D.D. in 1602.[56] His reputation was such that he was appointed one of the translators of the New Testament in 1604.[57]

A Medical Fellow following Gwynn was Richard Andrewes (1575–1634), who held the fellowship from 1605–1612. Andrewes came to Oxford in 1591, a year before Fludd, and held the rhetoric readership from 1603 to 1605.[58] Fludd and Andrewes were also colleagues in the College of Physicians in London, and Fludd called his contemporary his "worthy friend."[59]

Considering the fact that there were opportunities at St John's College for the study of medicine, unlike most other colleges at Oxford, and that the theologian-president had been a medical student, it becomes clear how Fludd could have been sufficiently exposed to the profession to decide on it as a career. It is also relatively easy to see that because of the nature of Fludd's humanist Arts education in an institution which still had as its primary aim the preparation of scholars for holy orders, his works all have a firm basis in Christian or pseudo-Christian teachings. Fludd makes frequent and often lengthy citations of Scriptural passages for use as the foundations for his scientific and philosophical arguments, exactly what one would expect from his education. He could just as easily have taken a doctorate of divinity and pursued a career in the Church of England, and would have felt altogether at home. Fludd counted among his most esteemed friends clergymen in high office, including the Bishop of Lincoln, the Bishop of Worcester and an Archbishop of Canterbury.

It must also have been at St John's that the small-statured, serious-minded Fludd began his compilation and organization of materials pertinent to his interests, materials which would appear in print about twenty years later. There was an influence present at St John's which would have provided one of Fludd's anti-

Aristotelian sources, as well as his organizational method: the work of Petrus Ramus (1515–1572) and his followers. There is no question that Ramism was strong at Oxford and at St John's before Fludd came and after he left. John Case (1546–1600), an M.D. and one-time Fellow of St John's, who taught in his home at Oxford after losing his fellowship, reflects on the interest in Ramus in 1585:

> Still I cannot but acknowledge that the youthful ardor of mind in both universities has of late been fighting it out to determine whether in the mastering of the arts the great acuteness of Aristotle is of more worth than the flowing genius of Ramus. But, as I expect, the young exalt such apostasy from the true experience of age and from the wise old custom of philosophizing, because beardless youth often does what white hair denies to have been rightly done . . . I do not blame Ramus in this, for he was learned; I rather exalt Aristotle, for he stands out above all. But perhaps the young men will hold my work the poorer because I name in it the old interpreters of Aristotle.[60]

Indeed, since the mid-1570s a number of Ramus' works had been available in England, including his *Dialecticae Libri Duo*, which was published in London in both Latin and English editions.[61] The 1613 inventory of the library of John English, a Fellow of St John's, whose books were in his rooms at the College, showed that he had four works by Ramus, including both of the London editions, and one by a disciple.[62] Fludd quotes one of the dedicated followers of Ramus, Johannes Thomas Freigius (one of whose works was in English's collection), and calls him a "very learned Naturalist, and a man who hath taken great pains in searching out the truth of natural mysteries . . . "[63]

It has usually been assumed that Fludd acquired his taste for Renaissance occultism and Paracelsianism after he left St John's College and travelled on the Continent, but it would appear that the opposite is the case. Frances Yates and Peter Ammann have already drawn attention to the passage in one of Fludd's books where he relates the story of being interrupted in his room at St John's by his tutor, Perrin, while working on his treatise on music. As Fludd put it, Perrin wanted him to use "mea Astrologia" to discover the thief who had just robbed the tutor, a favor which the studious esquire set out to do.[64] This story has some important implications for understanding Fludd's development. The first is that he was already considered a knowledgeable astrological practitioner. This must have been sometime during the years 1596 to 1598, because Fludd says that he had his B.A. by then. Second, his learned divine tutor did not think it amiss to ask him to perform such a task. Third, Fludd was already at work at St John's writing treatises following the Neoplatonic and Hermetic lore of the divine harmonies and the microcosm-macrocosm correspondences. It is apparent from this that Fludd had nurtured his Renaissance occultist views at Oxford, obviously not over his tutor's objections. This would also mean that they were not religiously objectionable, and that Fludd had set to work to organize his occult interests systematically.[65]

Thus when Fludd took his M.A. In July, 1598, and announced in the register that he was "going over sea,"[66] he already knew what he was looking for before he left. In other words, he intended to seek on the Continent more of what he had been studying at Oxford, a goal which would naturally attract him to Paracelsus and other Neoplatonists there. Exactly when he departed from St John's to travel is not known, but apparently he kept his room under the new-built library until Lady Day, 1600.[67] In his treatise *De Naturae Simia*, Fludd says that he spent nearly six years travelling "all through" France, Spain, Italy and Germany.[68]

In each of these countries Fludd undoubtedly sought out men and books in the universities and princely courts with views compatible with his own, the more to add to his collection of Renaissance Neoplatonic and Hermetic learning.

From England, Fludd probably travelled to Paris, and from there he stopped at Lyons.[69] On his way to Italy, snow blocking the St Bernard Pass through the Alps forced Fludd to spend the winter of 1601/2 in Avignon, from where he was summoned to Marseilles to instruct the Duke of Guise and his brother, a Knight of Malta, in mathematics.[70] The Oxford master travelled in Provence, then into Italy, where he says that he spoke to a merchant in Livorno who had just come from "Fess in Barbary."[71] In Rome, he met a master Gruter of Swiss birth, from whom Fludd claims he learned a great deal.[72] Last on his trip he mentions travelling "from Venice unto Augusta, or Augsburg, in Germany."[73] Among other places in Germany, Fludd very likely visited Heidelberg, seat of the university and the court of the Calvinist Elector Palatine, the future son-in-law of James I.[74] It is also highly possible that Fludd visited the court of Moritz the Learned, Landgrave of Hesse, whose sometime personal physician was Michael Maier, Fludd's philosophical brother, from whose presses came the later Rosicrucian manifestos.[75] After touring Germany, Fludd probably returned to England. His education was finished, and he only needed to return to pick up his medical degree and begin practice, while continuing to organize all the material he had gathered on the Continent.

FLUDD, MEDICINE AND CHEMISTRY IN EARLY STUART ENGLAND

De Medicine etiam perfectione parum gloriari possunt quibus
philosophiae Naturalis thesaurus occultatur, quoniam
philosophia naturalis est medicine basis seu
fundamentum . . . Nam cui mystica physices arcana incognita
sunt, hunc etiam occulta Medicinae secreta latere necesse est.

Fludd, *Tractatus Apologeticus*, pp. 89–90.

When he returned from his Continental journeys, Fludd entered
Christ Church College, Oxford, to take his degree in medicine.
Since there was an outbreak of the plague at Oxford in the late
summer of 1604,[1] Fludd probably did not re-enter Oxford until late
1604 or early 1605. The university register shows that on 14 May
1605 Fludd made his supplication, and on 16 May was granted the
M.B. and M.D. and licensed to practice medicine.[2]

The brevity of his stay before being granted the medical degree
was not extraordinary, since the main requirement for the degree
was proof that the supplicant had read the required medical texts,
primarily Galen and Hippocrates.[3] As part of his degree requirement, Fludd defended three theses:
1) "Frequent use of purgative medicines does not accelerate aging"
2) "Chemical extractions bring less harm and danger than whole
and natural ones" and
3) "The aged more easily bear fasting than the young."[4]

From Oxford, Fludd decided to set up medical practice in
London, for which an additional license to practice was necessary
from the College of Physicians, the governing body for the London
metropolitan area. Accordingly, six months after receiving his
M.D., Fludd applied to the College and was first examined on
8 November 1605, as reported in its *Annals*:

> On this day Dr. Fludd of Oxford applied for admission as a
> Licentiate, but when he was examined in both Galenical and
> Spagyrical medicines he was not satisfactory enough in either.

They therefore advised him to apply himself more diligently to his studies.[5]

Because of the results of this first examination, Fludd was directed to "abstain from practice."[6]

There are several points to be noted about this first examination before the College. The first is that it was not unusual for applicants for a license to be rejected on their first appearance before the Censors of the College; this was particularly true of those who graduated from the English universities. The best medical education was considered to be that obtained on the Continent, most notably at Padua, Bologna, Leyden and Basel.[7] Most of the members of the College had received their degrees on the Continent, as noted above. As a result, it is not particularly surprising that they found Fludd deficient in Galenical medicine. The second interesting fact is that Fludd was examined in "Spagyrical" or chemical medicine, and found deficient in this as well. Presumably this subject accounted for a great deal of Fludd's medical studies in his travels on the Continent, but it was likely to have been of a kind that was different from the chemical medicine of the Fellows of the College. It is clear from his work at St John's and his later publications that Fludd was interested in the Neoplatonic, Paracelsian cosmology as the basis for his chemical medicine, as was in widespread favor on the Continent. As Allen Debus has pointed out, in England, a limited, empirical approach to the use of chemical medicines was adopted without the Paracelsian cosmology, and this empirical chemistry was then combined with traditional Galenical medicine.[8] Thus it is not surprising that the College would not find Fludd altogether satisfactory in chemical medicine either.

Fludd must have applied himself in a way satisfactory to the College because, on his second appearance, on 7 February 1606, he was granted a license:

> Second examination, and even though his examination is not wholly satisfactory, nevertheless in the judgment of everyone he appears to be not uneducated, and permission is therefore given to practice medicine.[9]

Fludd had no sooner begun to set up practice than he seems to have run into trouble for extolling the virtues of his brand of chemical medicine. He was secretly denounced to the College of Physicians, as described in this entry in the *Annals* on 2 May 1606:

> There is information come to the College that Dr Fludd had praised himself and his chemical medicines a great deal and rejected Galenical doctors with contempt. Therefore the Censors ordered him to be summoned on this day. When he was questioned to the truth of this charge, he most confidently denied everything, and demanded to confront his accusers; since they did not appear, he was dismissed with the warning that he was to feel and speak modestly about himself and was to have respect for the Fellows of the College. And when he had fully discharged his obligations according to the prescribed statutes, he would be admitted into the number of those permitted to practice.[10]

Apparently Fludd's license was withdrawn on the above date, but he continued to practice and waited until the following year to apply as a Candidate for membership as a Fellow of the College, an honor which normally required several examinations and a probationary period. Fludd was examined for candidature on 1 August 1607, 9 October and 22 December of the same year, and on the latter date was judged to be worthy of being made a Candidate.[11] Unfortunately, at his next examination, on 21 March 1608, one of the Censors must have touched a sore spot with Fludd, probably by attacking some of his fondest cosmological medical beliefs. The *Annals* of that date tells of a less than congenial meeting:

> Dr. Fludd, who already had been selected to be numbered among the Candidates, conducted himself so insolently that everyone was offended, and was therefore rejected by the President with the warning that if he is to continue to practice without a license he should take great care.[12]

It would seem that the College was well aware that Fludd had been practicing medicine his own way without a license since May, 1606, but since he was obviously educated to the standards of the day and not a back-alley quack, they were perfectly willing to let it go as long as he did not additionally cast aspersions upon their own traditional medicine.

Eventually, cooler heads seem to have prevailed: just three months later, on 25 June 1608, Fludd was re-admitted as a

Candidate and admitted as a Fellow on 20 September 1609. In contrast to what Munk said, and many others have repeated after him (see Ch. X), Fludd became a distinguished member of the College. He served four terms as Censor (1618, 1627, 1633 and 1634).[13] In 1616, he participated in the inspection of the London apothecaries and afterwards gave a dinner at the College for all the Fellows and their wives.[14] Another example of Fludd's status as a Fellow in good standing, well after the appearance of his most controversial publications, was the fact that he gave the public anatomy lecture at the College on at least one occasion (27 June 1620).[15] Fludd was elected a Censor for the second time in 1627. That same year, the Privy Council ordered that the President of the College along with six other doctors and some London aldermen should look into a complaint about noxious fumes from an alum works in the parish of St Catherine's. Fludd was one of the examiners of the works, and the subsequent report to the Privy Council was signed by John Argent, President; John Gifford, the treasurer; William Harvey; Fellows William Clement, Fludd, Ottuell Meverell (also a 1627 Censor), Sir Simon Baskerville (physician to James I and Charles I); and an alderman, Hugh Hamersley.[16] Fludd was twice more chosen Censor, in 1633 and 1634. In 1635 he gave the College a set of his works, which were to be clasped and placed in the library.[17] There can be no doubt that Fludd was well received by the other Fellows of the College, including his good friends William Harvey and Richard Andrewes, as well as Sir William Paddy, to whom Fludd dedicated his *Medicina Catholica* of 1629. Membership in the College was also a matter of pride for Fludd: in addition to a set of his works, he left the College twenty pounds in his will, and he was probably the first member to mention his Fellowship on his church monument.[18]

Despite his philosophical speculations, Fludd was considered an established, orthodox medical practitioner, both by his colleagues and by the critics of the College in the mid-seventeenth century. It has been regularly assumed by later writers that Fludd's affinity for Renaissance religious Hermeticism led him to be little more than a witch doctor at best, or a foolish babbler of philosophical nonsense at worst, but nothing could be further from the truth.

One could give as another example of his medical orthodoxy the following. As a follow-up to the inspection of the London apothecaries in 1614, the College of Physicians issued its long-contemplated directory of standard pharmaceutical preparations in 1618, which was to be followed by all the apothecaries in the London area. Fludd's name appears at the beginning as one of the Fellows of the College and thereby an author of the work.[19] The *Pharmacopoeia Londinensis* went through at least twenty-three editions in the seventeenth century, and during the 1630s Fludd had a number of disputes with the apothecaries when he enforced the code as a Censor of the College.[20]

In fact, Fludd was considered to be such an establishment figure by Nicholas Culpepper and his follower Peter Cole in the mid-seventeenth century that he was lumped in with the rest of the College in their criticism of it. Cole published an English translation of the *Pharmacopoeia* in 1649 in accordance with his desire to break the monopoly of the College and make medicines more available to the layman.[21] Peter Cole also provides some interesting posthumous hearsay about Fludd. It is worth noting here because of Cole's inclusion of Fludd in the conservative College and what I take to be the truth about Fludd's practice of medicine. In publishing some of Culpepper's remaining manuscripts after his death, Cole adds a composition of his own called *Mr. Culpepper's Ghost*, wherein the famous herbalist is supposedly speaking from the Elysian Fields for the purpose of "giving seasonable advice to the lovers of his writings."[22] His main advice is against the restoration of the monarchy, which would re-install a "Luxurious and debauched court," and, just as bad, the "Monopolizing upstart London Colledg will be called to secret Court-Ministries, and . . . dare without much blushing to beseech their Debauched and Frenchified Lordships to become earnest soliciters to the King's most Excellent Majestie . . . ". Then would follow a College-instituted order by the King for the burning of all "Books of Physick in the English Tongue" by the "Hand of the Hangman at the places of such Martyrdoms, but especially at the Royal Exchange, over against Pestilent Peters own shop . . . " This would be done for the purpose of retaining a monopoly over physic and pharmacy by the select few.[23]

Then Cole implies that even though Fludd professed to be a Hermetic Paracelsian, he was really a Galenist in practice. The shade of Culpepper says that while he always had a high regard for chemistry, and still considers it useful for the search into the mysteries of nature as well as the best key to natural philosophy, experience has confirmed for him that Galenical and Hippocratical remedies are probably better. He continues:

> The occasion that made me first incline to this opinion was thus.
> I met here in the *Elysian Fields* accidently *Factor Wright junior*,
> who first lived with *Dr. Flud* in Fan-Church street, being a Youth
> and give out his Physick (as his House-Apothecary) who told me
> that his Mr. *Flud*, though a
> Trismegistion-Platonick-Rosy-crucian Doctor, gave his Patients
> the same kind of *Galenical Medicaments*, which other Physitians in
> the Town ordinarily appointed, and when himself was sick, he
> had no Chymical *Elixirs* or *Quintessential Extracts* to relie upon,
> but after he had caused himself to be let blood (an ordinary
> Galenical Remedy) he sent for Doctor Gulstone,[24] & Relied
> upon his advise for the Cure of his disease, who was a *pure
> Galenist*. And I wel remember in one of his Folio Books treating
> of the Preservation of Health, his only Physick he recommends,
> is a *Mastich Pill*, with a drop of *Oyl of Time*, than which the
> Apothecaries Shop knows no more common Medicine. And
> therefore he being so great, and so sublime a *Doctor*, as to have
> written many Volumes in Folio, full of *Mathematical, Mystagogical,
> Chymical-Rosicrucean Speculations*, which Books are highly esteemed
> by many beyond Seas, and by some at home. If all his Skill in
> *Chymistry* (of which the *Rosie Crucian Seraphical illuminated Fraternity*
> are the chief Masters) or other Mysterious Arts, had furnished
> him with many more effectual Medicaments than the Galenical,
> such as are in the *London Dispensatory*, I cannot doubt but he
> would have used them, if not for Conscience sake, yet for to
> advance his Reputation, by the *quickness, safeness* and *pleasantness*
> of the Cure, which is that the Chymists boast of.
>   After the foresaid relation of *Factor Wright junior*, I happened to
> meet with *Old Dr Flud*, walking very musefully, and
> communicating with the Ghosts of *Raimundus Lullius*, and *Van
> Helmont*; I took old Dr. *Flud* aside, and asked him if the Relation
> of his Servant *Robert Wright* were true. Hereupon the old man
> ingenuously confessed the thing, and said there were no better
> Medicaments than the Galenical . . . [25]

Whether or not Fludd's former servant, Robert Wright, did indeed impart this information directly to Culpepper, we do not know; but

I believe that the above passage reflects a good deal of truth about Fludd's actual medical practice.

Allen Debus has shown the compromise which was reached in England between Paracelsian chemical medicine and the traditional medicine of Galen and Hippocrates. In this "Elizabethan compromise . . . chemical medicines which proved useful in combating disease were . . . accepted . . . [while] on the other hand . . . the Paracelsian mystical universe . . . became an object of distrust and suspicion."[26] There were others, however, as Debus points out, who were "interested in the theoretical work of Paracelsus [and] often paid little attention to his practical medical reforms."[27] If we can believe the passage from Cole's work above, this appears to be the case with Fludd. His metaphysical speculations were a consuming interest, but it is apparent that they did not carry over into a radically different medical practice.

Fludd treats the metaphysical mysteries behind all of the applied arts, including medicine, in every one of his works, and insists that those who do not properly understand the unchanging truth of the arcana behind the ever-changing world of physical phenomena (the Aristotelians and Galenists) have a false natural philosophy and thus at best a haphazard medical practice based upon it. The basis of Fludd's metaphysical theories (discussed in detail in Chapter VIII) is a Neoplatonic interpretation of the Old Testament, including the Apocrypha, bolstered by Ficino's translations of Plato, the Neoplatonists and Hermes Trismegistus. He also makes great use of Agrippa's *Occult Philosophy*, Reuchlin's *Cabala* and other occult and alchemical collections.

According to Fludd's history, disease is caused by a wind controlled by an evil spirit which excites the ever-present lesser evil spirits in the air (Plate 11). These unwanted diabolical servants enter the body through the pores or from breathing. If there is a weakness in a bodily organ, the evil forces attack the bonds which hold the four humors in balance, causing a malady; if this is not corrected, the imbalance will become so violent it will cause the entire physical "palace" to collapse. Treatment of this imbalance is through administration of remedies to counteract the evil wind by providing, through sympathy, those angelic forces which were lacking and caused weakness in the first place (see Plates 12–13). This sympathetic natural magic could be administered through the use of proper herbs, chemical medicines or "magnetic" preparations designed to work from a distance. Fludd's medical theories were an elaborate synthesis of ancient, medieval and Renaissance learning and contemporary experimental thought.[28]

In this chapter we are considering how this theory was applied in practice. It would appear that although Fludd's metaphysics differed greatly from that of the dyed-in-the-wool Galenists, his actual medical practice, with few exceptions, did not. As a member of the College of Physicians, he was nominally an author of the *Pharmacopoeia Londinensis*; he kept an apothecary at his own house and was, when Censor of the College, an enforcer of the standard pharmacopoeia. From this it is obvious that Fludd prescribed the same herbal and chemical remedies as every other orthodox physician. He cast the horoscope (as Hippocrates insisted), examined the urine, took the pulse and listened to the complaint of the patient, all of which were part of standard practice.[29] The only likely difference in Fludd's practice was the use of "magnetical" cures working from a distance, such as the "weapon salve," sending the urine to the countryside, or the making of a "magnet" to be worn on the body to draw out an affliction; but even in this case there were some other physicians in England and more on the Continent who believed in and made use of these cures.[30]

There is additional testimony for the orthodoxy of Fludd's practice as well as the "modernity" of his medical research. With his friend and colleague Dr Richard Andrewes, Fludd joined the Barber-Surgeons company, presumably in order to practice surgery without criticism.[31] As mentioned before, Fludd gave the public anatomy lecture to the College at least once, and had often observed his colleague William Harvey's dissections of the heart, which led to the formulation of the theory of the circulation of the blood.[32] However, Fludd did not simply watch others: he also carried out his own anatomical experiments. Harvey quotes pathological experiments made by Fludd and others in his notes,[33] and Fludd himself mentions an experimental splenectomy he performed on a dog.[34] Furthermore, Fludd very much believed in attempting experiments to explain medical pathology as well as natural philosophy. Even though it is true that his experiments were done in such a way that they appeared to confirm his *a priori* cosmology, the important fact here is that he *did* experiment, and extensively.[35]

When the experimentation became a little too esoteric, perhaps involving higher forces that he was not sure he could control,

Fludd apparently backed away, however. In his experiment on the properties of wheat, he discovered a miraculous balm, but never used it again:

> . . . verily I must really confess that I once finding my selfe very much molested with an ach and debility in the back of mine own hand, and being at that very tyme occupied about this very same experiment which I handle at this present, I took a little of that crude quintessential balme of wheat and therewith did annoynt the back of my hand, & from that hower being three yeares since till this very season I never felt that payne wherwith I was ordinarily affected before that tyme. And yet I profess (with shame be it spoken) eyther my negligence or other occasions have beene such y^t neyther before nore since did I ever make any further triall of the property thereof. For as much as in my spirit I have wished that by thos experimented philosophers who have been better practiced and acquainted wth the secret vertues and Nature of this first Element, I might be fully and truly taught a way to exalt it to his uttermost effect of purity.[36]

It has always been assumed that all of Fludd's experimentation resulted in nothing of practical use, not only from a "modern" view, but also in the opinion of the seventeenth century. This is not true. Fludd apparently set up a sizable chemical laboratory and brought in a French technician to run it, with one very practical result: the development of a higher quality steel than was currently being produced in England. On 12 May 1618, the two holders of the royal monopoly on steel-making in England complained to the Privy Council that Fludd had brought in a Frenchman, had set up furnaces, and was encroaching on their monopoly by making steel.[37] On 30 May 1620, James I ordered the Council to consider Fludd's counter-petition asking for a patent to make steel using his "forraine" operators, "with the purpose to imploy and sett them on worke here in that mystery, for the benefitt and behoofe of the publicke." Their Lordships allowed that current steel-making was insufficient in quality, and that "the makeinge of good and serviceable steel within the kingdome is a matter of good consequence, and worthie of all due encouragement."[38] At a meeting of the Council on 27 September 1620, which included the King, the Archbishop of Canterbury, the Lord Chancellor Francis Bacon and the other members, Fludd was granted a patent:

... forasmuch as Robert Fludd, doctor of phisick, hath, at great charge, drawen over hither from forraine parts certaine persons, and amongst others one John Rochier, a Frenchman, skillfull and expert in makeinge of steele, and, being an humble suitor for a patent to sett them on worke in that mystery for the good of the publicke, did this day offerr to the Board a certificate, under the hands of manie cutlers, blacksmyths, and other artificers workinge in steele, that the steele made by the said Rochier is very serviceable, good and sufficient, and Doctor Fludd further undertakinge to make a good steele as anie is made in forraine parts, and to vent the same at easier and cheaper rates than the outlandish steele; that they will waste noe wood but only make it of pitt coale; that they desire noe barr of importation more than what the goodness and cheapness of their stuffe shall occasion, and thereby that his Majestie shall have a third part of the profitt arryseinge thereby ... [39]

Thus not only did Fludd offer to make a higher quality steel with no waste of wood, he sweetened the bargain by proffering one-third of the profits to James, an offer which no doubt helped secure the patent at a time when the monarchy frequently sought to derive revenue from the granting of monopolies.

There also can be no doubt that Fludd was a respected medical practitioner and experimenter in his day. The iatrochemist Daniel Sennert, who disagreed with Fludd about the weapon-salve cure, wrote in a publication of 1637 that Fludd was wrongly abused by Parson Foster, even though he had truth on his side, because Fludd "is a learned Doctor, well esteemed at home for this practicall skill in Physicke, and much honored abroad for his learned books in Print."[40]

The conclusion we are drawn to is exactly as Sennert expressed it. Fludd's metaphysical speculations, while honored on the Continent, were mostly ignored in England; at home, however, he was a well-respected medical doctor and chemical experimenter who, to a great extent, employed conventional Galenical remedies in his practice. Apparently he was fond of lecturing his patients on his metaphysics, a habit which gave him the reputation of using "chants" to work psychological cures. It would seem also that it was his Hermetic philosophy which involved him in ceaseless controversy and, in the end, overshadowed his contemporary reputation as a successful physician and practical experimenter.

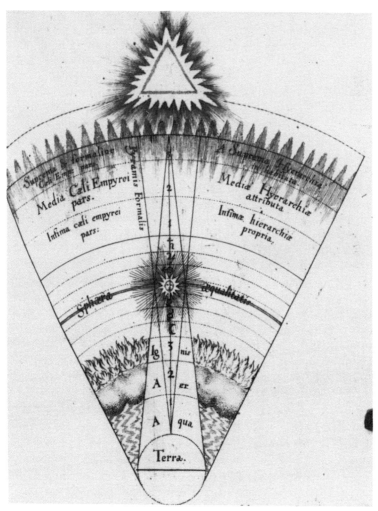

2. The Three Heavens and Inverted Pyramids of Form and
Matter (UCH I, 1, p. 89)

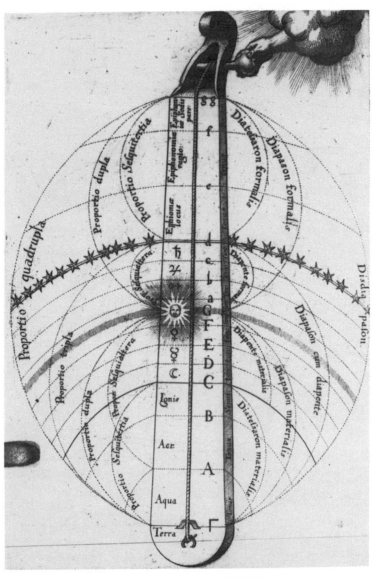

3. The Celestial Monochord (UCH I, 1, p. 90)

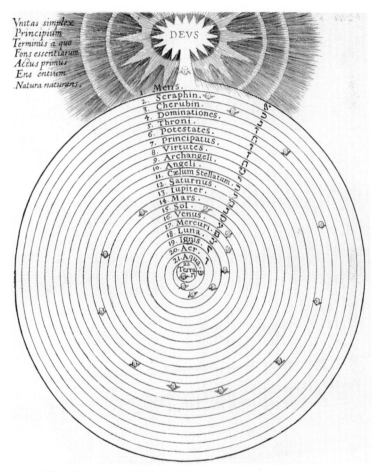

4. The Downward Spiral of Creation (UCH II, 1, p. 219)

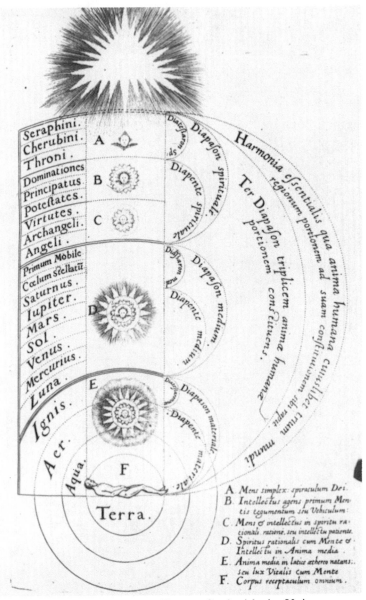

Seraphini.
Cherubini.
Throni.
Dominationes
Principatus
Potestates.
Virtutes.
Archangeli.
Angeli.
Primum Mobile
Cœlum Stellatū
Saturnus.
Iupiter.
Mars.
Sol.
Venus.
Mercurius.
Luna.
Ignis.
Aer.
Aqua.
Terra.

A
B
C
D
E
F

Harmonia essentialis qua anima humana cuiuslibet trium ibi rapit regionum portionem ad suam constituendam
Ter Diapason triplicem animæ humanæ portionem constituens.
Diapason spirituale.
Diapente spirituale.
Diapason medium.
Diapente medium.
Diapason materiale.
Diapente materiale.
Diatessaron sp.
Diatessaron med.
Dia. materiale.
ipsum mundi

A. *Mens simplex: spiraculum Dei.*
B. *Intellectus agens primum Mentis tegumentum seu Vehiculum:*
C. *Mens & intellectus in spiritu rationali ratione, seu intellectu patiente.*
D. *Spiritus rationalis cum Mente & Intellectu in Anima media.*
E. *Anima media in lotus æthereo natans: seu lux Vitalis cum Mente.*
F. *Corpus receptaculum omnium.*

5. The Harmony of the Human Soul with the Universe
(UCH II, 1, i, p. 93)

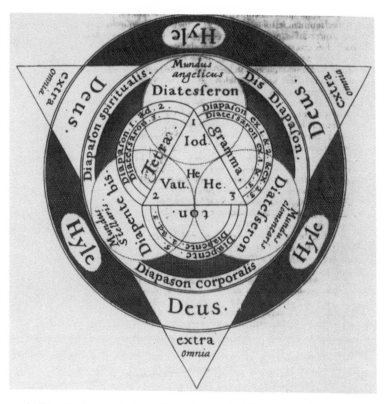

6. The Emblematic Manifestation of the Trinitarian Nature of
the Universe (UCH II, 1, i, p. 92)

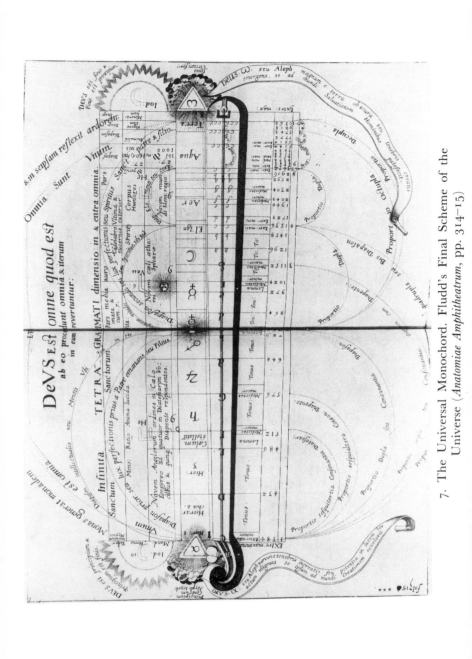

7. The Universal Monochord. Fludd's Final Scheme of the Universe (*Anatomiae Amphitheatrum*, pp. 314–15)

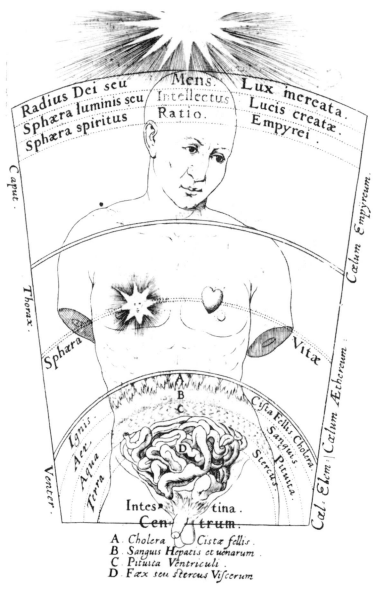

Radius Dei seu
Sphæra luminis seu
Sphæra spiritus

Mens.
Intellectus
Ratio.

Lux increata.
Lucis creatæ.
Empyrei.

Caput.

Cælum Empyreum.

Thorax.

Cæl. Elem| Cælum Æthereum :

Sphæra

Vitæ

A
B
C
D

Cista Fellis Cholera
Sanguis
Pituita
Stercus

Ignis.
Aer.
Aqua
Terra

Venter.

Intes—   —tina.
Cen—   —trum.

A. *Cholera*   *Cistæ fellis.*
B. *Sanguis Hepatis et uenarum.*
C. *Pituita Ventriculi.*
D. *Fæx seu stercus Viscerum.*

8. Microcosm-Macrocosm Related to the Three Heavens
(UCH II, 1, p. 105)

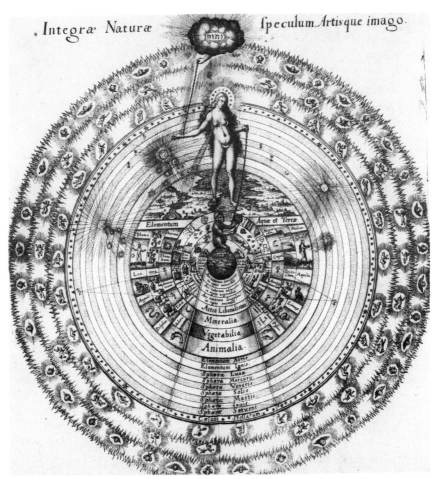

9. Universal Nature, Man (the ape of Nature) and the Different Regions (UCH I, 1, i, pp. 4–5)

FRIENDS, ASSOCIATES AND PATRONS

... there is no surer sign of a man's moral character than this, whether he is or is not destitute of [tried and true] friends.

Plato, Letter VII, 332c

Wise men say that the heavens and the earth, gods and men, are bound together by fellowship and friendship, and order and temperance and justice ...

Plato, *Gorgias*, 508a

Until now, discussions of Robert Fludd have centered on aspects of his written works and the controversies which they generated in England and the Continent. What has been missing is the personal context of the circles of his friends and associates which would help illuminate his involvement in the social, intellectual and professional life of his day, and give us a more integrated picture. While examination of his works and some contemporary reaction to them is certainly important, it can present a point of view which is easily subject to distortion. This is particularly true, of course, if his works are isolated from their context and compared to modern, "correct," (at least for the moment) mechanistic science, which was beginning to emerge as dominant in the seventeenth century. Therefore a brief survey of Fludd's friends, associates and patrons may serve to illustrate that he was accepted by many as a legitimate spokesman for another valid world-view of the time, that of Renaissance Christian Neoplatonism stemming from Florence. It would also show that Fludd and his works were not an isolated aberration, but that he moved among many notable figures who either shared, or did not disdain his views.

The first place to look for lifelong friends and acquaintances of Robert Fludd is late Elizabethan Oxford. From their days together at St John's, Fludd's closest College friend was Richard Andrewes (1575–1634). As noted in Chapter I, Andrewes entered St John's one year before Fludd, in 1591, at the age of sixteen.[1] He received his B.A. in the usual four years, but did not take his M.A. until 1599.[2] From 1603 to October 1605, Andrewes was the lecturer in rhetoric, but resigned to practice medicine for Charles Blount, Lord Mountjoy (after 1603 the Earl of Devonshire).[3] On Mountjoy's death in April 1606, Andrewes must have returned to Oxford to resume his studies as Medical Fellow; he had made supplication for the M.B. degree in February 1605, but did not receive it until 1608, along with his M.D.[4] Two months after Fludd was admitted as a Fellow to the College of Physicians in 1609, Andrewes was accepted as a Candidate, and became a Fellow the following year.[5] The College elected Andrewes to be Censor five times, and his medical skill was such that in 1631 he was appointed to succeed William Harvey as physician to St Bartholomew's Hospital.[6] As previously noted, Fludd considered his old Oxford fellow-student his "worthy friend."[7]

Besides his tutor and the President of St John's already discussed, Fludd must have known, but apparently was not a close friend of, William Laud, the ill-fated Archbishop of Canterbury. Laud was a year older than Fludd, a pupil of St John's who received his B.A. from Oxford the year before the esquire from Kent, but they both became M.A. in 1598, only twelve days apart.[8] On 3 September 1603, Laud entered the service of Lord Mountjoy (two months previously made Earl of Devonshire by James I for his service in Ireland) as his chaplain, and it may have been through Laud's recommendation that Richard Andrewes became physician to Mountjoy's household.[9] Known for his strong anti-Puritan, High Anglican views, Laud was President of St John's from 1611 to 1621. With his predecessor and tutor, John Buckeridge (c. 1562–1631), and successor, William Juxon (1582–1663), Archbishop of Canterbury 1661–1663, they formed a dominant High Anglican party at St John's and Oxford.[10]

It therefore seems significant for an understanding of Fludd's religious views to note that while there can be no question that Fludd knew Laud and Buckeridge, the London doctor dedicated two of his works to arch-enemies of Laud, George Abbot (Archbishop of Canterbury from 1611 to 1633) and John Williams, Bishop of Lincoln, Archbishop of York, and Lord Keeper of the Great Seal following Francis Bacon.[11] Abbot was a strict Puritan reformer and had been an enemy of Laud's since 1603. Abbot, at that time Vice-Chancellor, had heard a divinity lecture given by the St John's Fellow and took great exception to Laud's statement that "the Church of Christ derived from the apostles and the Church

26

of Rome, continued in that Church . . . to the Reformation."[12] Abbot opposed Laud's election to the presidency of St John's, Oxford, in 1611, but James, after a three-hour hearing, decided that the close election was valid and accepted Laud as President.[13] Going so far as to write a counter-blast to Laud's views that was circulated in manuscript and published in 1624, Abbot remained a life-long foe of Laud's.[14] Although it is not known whether Fludd knew George Abbot personally, he undoubtedly was familiar with Abbot's theological views. While Fludd was at Oxford, Abbot was drawing large crowds to his sermons at St Mary's, and six theses which Abbot publicly defended in 1597 (probably for his D.D.), were published in 1598 and re-published by Abraham Scultetus at Frankfurt in 1616.[15] There was never any doubt in the mind of Fludd or anyone else in England where Archbishop Abbot stood religiously. So, the dedication of a book to him clearly shows that Fludd was essentially in agreement with the Archbishop's broadly based Protestantism, particularly since Abbot had been at Canterbury for twenty years by the time of the dedication.

This is bolstered by Fludd's earlier dedication to Laud's closest rival for royal favor, John Williams. Williams was a religious moderate, which is another way of saying that for him religion was a sideline, and that church offices were the means by which he could acquire wealth, influence and power. Though he never shrank from using reference to divine authority, his religious position was normally that which he felt would keep him in favor at the moment, but of importance here is the point that he disliked religious extremes on both sides, and remained Laud's constant enemy (even to the point of giving Laud bad dreams).[16]

When Fludd was accused of heresy and religious innovation in his printed apology for the Rosicrucians, he offers this explanation to King James:

My *Tractatus Apologeticus* does not deal with some religious innovation, nor does it share even an iota of any heresy, inasmuch as I, the author of that work, . . . have upheld an adherance to this reformed religion now the custom among us . . . which I received from my infancy, in fact from the time I lay at the breast of my nurse in England . . . to this day.[17]

In a later document, he invokes higher authority:

27

Judg therefore and witnes in the first place yea heavens and that
everliving power whose tabernacle ye ar [i.e., mankind], if I ever
knew or contracted my wishes unto any Religion, saving this
reformed one, so happely celebrated heare in England . . . [18]

From this it would appear that Fludd was a dedicated Anglican,
definitely more of a Calvinist than a High Anglican, perhaps even
favoring the strict Puritanism of Abbot, since he remained
unmarried and insists, in 1618–1619, when he was about forty-five,
that he has remained a *virgo immaculata*.[19]

On leaving Oxford sometime between 1598 and 1600, Fludd
travelled on the Continent until late 1604 or early 1605. There he
met a number of physicians, princes and other learned men with
whom he stayed in contact after his return. Apparently he even
made return trips from time to time for closer consultation with his
Continental colleagues, as we learn from the following informative
letter (the only extant letter of Fludd's known), in which he also
gives a brief description of himself; it is included with a letter
written by Sir Thomas Fludd to the Earl of Salisbury:

> 1606, August 20 – a Son of mine being a doctor of physic and
> greatly desirous to have conference with certain physicians,
> Italian and French, now in France, his acquaintance and good
> friends in his travel beyond seas, touching secrets and other
> things concerning that study, and to return within two months,
> about a month past went over into France and landed at Dieppe,
> from whence he wrote the letter herein . . .

It seems that Robert had stumbled upon a possible source of entry
into England by traitors; he had been stopped momentarily on the
English coast because he matched the description of a suspect, and
found out more from boatmen lodgers at his inn, the "Fleur de
Lewse," in Dieppe:

> Robert Fludd to Sir Thomas Fludd, his father. I was examined
> lying at Hide [Hythe] because I was of the description of him
> that was searched after, for he had, they said, a small stature,
> lean visage, auburn hair, etc. . . . [20]

During his travels on the Continent, Fludd met a variety of men,
some noble, some learned, and some having both attributes. There

28

seemed to be no religious barrier, because some were Calvinist and some were Roman Catholic.

In France, Fludd tells of hearing a weird alchemical tale from a Lord Menanton living in Paris.[21] The apparition arising from the application of various degrees of fire to some human blood in a laboratory was also witnessed by Lord Bourdalone, who became a friend of Fludd's. Bourdalone was the Chief Secretary to Charles of Lorraine, fourth Duke of Guise, and apparently corresponded with Fludd, who called the lord a "most learned man."[22] The French noble returned the compliment, as shown in the line from this letter he sent to the London doctor: "Bearing before you a fine example of your genius, learning, and dilligence, it is certain that you accomplish what is especially befitting a virtuous man and a philosopher . . ."[23]

It must have been not too long after he left England that Fludd met Bourdalone, for when the travelling English doctor was forced by snow blocking the passage to Italy to spend the winter of 1601–1602 in Avignon, he was summoned by the Duke to instruct him and his brother "in the mathematical sciences," at Marseilles.[24] The choice of the learned Englishman was not a random one. Bourdalone was interested in philosophical speculation and alchemy, but the Duke himself was a believer in occult divination. Ten years previously Guise had escaped imprisonment in Tours (imposed by Henry III after the King had had his guards assassinate Guise's father in 1588) by vowing a pilgrimage to Our Lady of Loretto in Italy and on the basis of *"chances cabalistiques"* which foretold August, 1591, as a good time to attempt escape.[25]

Fludd was now tutoring two members of a premier family of France. His elder pupil had been proposed by the Catholic League as the King of France to succeed Henry III, but Philip II of Spain wanted his daughter Isabella on the throne, and a marriage between them could not be worked out.[26] At the time of Fludd's sojourn, Guise was the Governor of Provence, an Admiral of the Levant and sometime companion of the King, Henry IV.[27] The other pupil was the Duke's brother, François Paris de Lorraine, known in his time as a swashbuckling Chevalier, described as *"le plus galant homme du monde, puisqu'il avait tué huit ou neuf hommes en duel,"* and who was complimented by the Regent for his success and bravery in battle.[28]

In the preface to the second treatise of his *History of the*

Macrocosm, Fludd says that he composed the section on arithmetic for the private instruction of the Duke, to whom he dedicates it. Similarly, the sections on geometry, perspective and the "secret military arts" were drawn up for Francois' benefit.[29]

The Duke's younger brother, Fludd remarks, was unusually swift in leaning, and even tried to overtake his instructor in practice.[30] This short preface to the "*Lector benevolus*" also contains some clues to other contacts the English Master of Arts had in France. He dedicates the sections on music and the art of memory to another noble student, the Marquis de Orizon, Viscomte de Cadenet.[31] The art of geomancy was written in 1602 for the Papal Vice-Legate in Avignon,[32] and Fludd gratefully acknowledges the composition of the sections on the art of motion and astrology to his "very dear friend," Reginaud of Avignon.[33] Considering his interests, it is also not hard to understand that Fludd speaks of going to the Papal Palace in Avignon to converse with the Vice-Legate in the company of "a very dear friend of mine," Monsieur Malceau, the papal apothecary.[34]

In Rome, Fludd says that he not only learned about motions and machines, but also the "magnets" used for curing at a distance:

> I was, whilest I did sojourne in Rome, acquainted with a very learned and skilfull personage, called Master Gruter, hee was by birth of Swisserland, and for his excellency in the Mathematik, and in the Art of motions and inventions of Machins, he was much esteemed by the Cardinal Saint George: This Gentleman taught mee the best of my skill in these practices . . . and among the rest, he delivered this magnetical experiment unto me, as a great secret, assuring me that it was tried in his Country, upon many with good successe.[35]

After sojourning in Italy, Fludd went into Germany, where he no doubt sought out centers of learning to exchange knowledge of particular interest to him. Among the places visited were probably the courts of Frederick IV (1574–1610), Count Palatine in Heidelberg, and of Moritz the Learned, Landgrave of Hesse-Cassel (1572–1632). The attractions in Heidelberg included its university and well-known library (destined to grow to even greater renown under Janus Gruterus), in addition to the warm relations between England and the Palatinate.[36] Moritz of Hesse-Cassel was a ruling prince one could truly call an *uomo universale* in the

Renaissance tradition. His devotion to and promotion of literature and the performing arts was widely known, but equally famous was his interest in philosophy and science. He set up a chemical laboratory, built the first permanent German theater, founded a school and welcomed the learned from all disciplines to his court.[37] Moritz was carrying on a tradition built up by his father, William IV the Wise (1532–1592). William was a skilled botanist, built his own astronomical observatory (Tycho Brahe was among his correspondents), had a chemical laboratory and was a devoted mathematician.[38] The senior Landgrave carried on extensive correspondence with Queen Elizabeth, and was visited by the famous English occultist John Dee at Cassel in 1584 or 1585.[39] The court of the Landgrave of Hesse-Cassel was a major center of learning in that part of Continental Europe with close ties to England,[40] one Fludd was not likely to miss.

Another tie between Landgrave Moritz and Fludd was the physician and fellow mystical philosopher Michael Maier. Maier had been the personal physician of Emperor Rudolf II, and, sometime after his patron's death in 1612, he travelled to England, where he probably visited Fludd and met Sir William Paddy, James I's physician, among others. It may have been Maier who carried Fludd's *History of the Macrocosm* to his publisher Theodore de Bry in Oppenheim, thus beginning the long string of Fludd's publications by that printing house.[41] By 1619, Maier was Moritz's personal physician, and continued to practice medicine in Hesse until his death in Magdeburg in 1622.[42]

The results of Fludd's travels on the Continent for his intellectual development and sustenance were great and long-lasting. At Oxford in the 1590s his interests had already been developed, and he began to sketch out some of them formally while still there.[43] On the Continent he was able to broaden and fill out much of his pre-determined philosophical framework and begin to organize it into a formal system and sub-systems, which would begin to appear in print in 1617, and would be consistently maintained until his death. Thus not only is it likely that he acquired much material relating to his interests through meeting the learned on the Continent, but the desire to organize his learning for the instruction of others through both text and illustrations has resulted in some of the most interesting books printed in the first three centuries after Gutenberg's press, and

31

some of the most controversial from that era to this.

In Stuart England before the Civil War, when he built a prosperous medical practice in London and published his numerous works, Fludd was associated with three distinct learned circles, and with the crown itself. The first circle comprised those interested in alchemy, complete with all of its Renaissance Hermetic-Christian metaphysics. Hartlebury Castle, the seat of John Thornborough (1551–1641), Bishop of Worcester, served as the center for alchemists during this period. In the dedication to a work published in 1623, an admirer of Thornborough's styled the castle as "an Apollinian retreat, as a living library, a flourishing Academy, or a religious Abbey."[44] Fludd was a good friend of the Bishop and his son, Sir Thomas Thornborough, and stayed at the castle on at least one occasion. In dedicating his *Anatomiae Amphitheatrum* of 1623 to Bishop Thornborough, Fludd calls the Bishop "my singular friend, most studious in accurately inquiring into the mysteries of nature, in whom is the true light of the world, and the treasure of treasures."[45] Thornborough was clearly in close agreement with Fludd's Hermetic-Christian metaphysics: the Bishop wrote a treatise on the philosopher's stone in 1621 containing four direct marginal references to Fludd's *Tractatus Theologo-Philosophicus* of 1617, and the textual references show a knowledge of and agreement with Fludd's statements.[46] The visit to the Bishop's castle is described thus:

> In the great sickness time [undoubtedly 1625], I came out of
> *Wales*, and remaining for a while with my noble friend, the Lord
> Bishop of *Worcester* at *Hartebury* Castle, there I was advertised of a
> strange mischance which happened by lightening and thunder,
> about five weeks before my coming thither, some three or four
> miles from the Castle. I would needs go see the place, and in the
> company of my worthy friends Mr. *Finch*, and Sir *Thomas
> Thornborow* . . . [47]

A curiosity about Bishop Thornborough which has not been completely assessed is that he employed the notorious Simon Forman (see Chapter X) as his servant while attending Magdalen College, Oxford. A cross-influence, and thus a link with Fludd, is a possibility.[48]

The second circle with which Fludd was associated included the scholars and antiquaries William Camden (1551–1623), John

Selden (1584–1654) and Sir Robert Bruce Cotton (1571–1623), which had at its center Cotton's renowned library at his house. Cotton had been a pupil of Camden's at Westminster School, where Camden, the school's second master, probably interested the young student in antiquarian pursuits.[49] Camden published his monumental history *Britannia* in 1586, and Sir Robert Cotton started his library about two years later. Around these two grouped others with historical and other interests. These meetings developed into regular discussions and then into a formal Society of Antiquaries about 1585, which flourished until about 1608.[50] Apart from the formal society, Cotton's library and its fame continued to grow: Francis Bacon and Ben Jonson were often there, and Camden and the young John Selden spent many days among the unusually large and excellent collection.[51] Both of the latter published historical works dedicated to Cotton and acknowledged the great value of his library for their studies. Another guest at Cotton's house, logical to some and surprising to others, was John Dee, who had amassed the first great Elizabethan library and antiquarian collection of artefacts.[52] According to Meric Casaubon, Sir Robert Cotton bought what remained of Dee's library after his death.[53]

Fludd's ties to both John Selden and Sir Robert Cotton are clear, and through the latter he must have had access to the remainder of John Dee's rich collection of Renaissance occultist texts. Selden was a noted legal scholar and antiquarian, who praised the "noble studies" of Reuchlin, the great cabalist scholar, and Roger Bacon, two of Fludd's admired authors.[54] He praised Fludd highly in the dedication (to a friend) of his widely-read *Titles of Honor*:

> Some yeer since it [this book] was finish't, wanting, only in some parts, my last hand; which was then prevented by my dangerous and tedious sicknesse; being thence freed (as you know too, that were a continuall, most friendly, and carefull witnesse) by the Bounteous humanitie and advise of that learned Phisician Doctor Robert Floyd (whom my Memory alwaies honors) I was at length made able to perfit it.[55]

Fludd dedicated the second treatise of his *Medicina Catholica* to Sir Robert Cotton, in which he is addressed as "amico meo singulari."[56] The value of these friendships for Fludd was that in addition to the books he undoubtedly brought back from his

Continental travels, he had access to one of the largest, if not *the* largest collection of books in England, a collection which must have included a good deal of material compatible with his own views. These associations also provide clear evidence that despite the controversies which his works generated with Kepler and the mechanists on the Continent (see Chapter V), he was a valued colleague in learned English circles, and that many in those circles obviously shared a number of his views. This appears to be true for the third learned circle to which he belonged, that of his medical colleagues.[57] It was pointed out in the previous chapter that Fludd was closely associated with some of the leading physicians of the day, such as Sir William Paddy (physician-in-ordinary to James I) and William Harvey, who, it has been suggested, may have derived his ideas on the circulation of the blood from Fludd's metaphysics.[58]

It is clear that Fludd also enjoyed the patronage of the first two Stuart monarchs, James I and Charles I, which is discussed more in Chapter IV. While Fludd's relationship with James can be pieced together with some certainty, our knowledge about the connections with Charles remains incomplete. James loved to discourse with men of letters concerning their own interests; three times he summoned John Selden to court for a dialogue about the history of tithes,[59] and apparently did the same with Fludd following the publication of the *History of the Macrocosm*, which was dedicated to the King.[60] James seemed to have questioned Fludd closely about his theories, and to have been completely satisfied with the answers.[61] For example, Fludd says that he discussed the elements found in wheat and bread with the King, and that " . . . it pleased his most Excellent and learned Ma[tie] to object unto me that these elements were not simple ones and that most rightly . . . but as I answered him . . . "[62]

After his meeting or meetings with the King, the evidence supports Fludd's claim that he subsequently enjoyed James' patronage "all the dayes of his life."[63] Just as Sir Thomas Fludd had always been welcome at the royal court, so too was his son Robert, and it did not end with James I. In 1629, Charles I granted Fludd and his heirs the rents from a country manor and lands in Suffolk.[64] Exactly why this was done remains unknown, but it is a distinct indication of the continued royal favor and patronage that Fludd enjoyed for nearly the last twenty years of his life.

The portrait that emerges from a look at Fludd's professional practice in medicine and chemistry, and his learned associations in England and the Continent, is very different from one that only looks at his published works and considers the debates they generated and their non-acceptance from a modernist point of view. He was an important intellectual figure and practitioner of the day, and accepted by many leading people of the time as such, a view that has substantial evidence to support it. It was only when the entire intellectual climate changed during the course of the seventeenth century and beyond that Fludd, who represented so well a vital part of the Renaissance heritage, was no longer considered important because his views did not conform to the ascending mechanist ones.

Chapter IV

ROYAL PATRONAGE: THE "DECLARATIO BREVIS" TO JAMES I

> . . . I have felt the . . . blast of envy in many of mine actions, yet
> have I secretly smiled with my self, and rejoiced to think, that
> mine inner conscience . . . doth solace and comfort me, in
> justifying herself to be altogether ignorant of mens slanderous
> taxations and imaginary reports.
>
> Fludd, "A Philosophical Key"

The glimpse of Fludd's intellectual and professional circles in
England outlined in the previous chapter provides evidence of the
widespread acceptance in those communities of Fludd as a valued
colleague. This clearly contradicts the view of Fludd as an
interesting anomaly, a clever but unimportant occultist outsider,
and also lends weight to the argument that his views were an
important step in the transition from medieval Aristotelianism to
modern mechanism during the Scientific Revolution. Some of the
best evidence for the acceptance of Fludd's ideas, however, comes
from one of the toughest critics of the age, King James I: after a
personal examination of Fludd's work during a royal audience, the
King thereafter extended his patronage, which was continued by
his ill-fated son, Charles I.

Fludd's "Declaratio Brevis" to James I is an unpublished
manuscript which was probably written about late 1618 or 1619.[1]
It appears to be a follow-up to his meeting with the King to defend
his recently published apology for the elusive Rosicrucians and the
History of the Macrocosm, the first portion of his *magnum opus*. The
background and circumstances surrounding the "Declaratio"
provide some of the most valuable insights into Fludd's place in his
own time that we presently have available.

Fludd began writing some of the treatises that made up his
History of the Macrocosm while still a student at Oxford in the 1590s.
Over the course of the following twenty years, he completed his
education, established his medical practice, set up an operating
chemical laboratory, and finished work on the *History*, but had not

found a publisher for it. In 1614 and 1615, some manifestos appeared in Germany, which claimed to be written by the Brothers of the Rosy Cross, or Rosicrucians.[2] They called for a general intellectual and spiritual reformation of Europe based on Reformed Christianity and the "true ancient wisdom." The tracts called for all those who believed in their principles to declare themselves in print, and they would be considered for inclusion in the Brotherhood. Many would-be adherents did hasten to the printers with laudatory works, but the Brothers were attacked by a prominent German iatrochemist, Andreas Libavius, in his *Analysis Confessionis Fraternitatis De Rosea Cruce* of 1615.

Libavius' book caused Fludd to come to their defense by dashing off a short apology, the *Apologia Compendiaria*, so that it could be listed in the 1616 Frankfurt book fair. In it, he answered Libavius' objections, praised the Brothers, outlined a longer apologetical work to follow, and appended a supplication for the Brothers to consider his worthiness for membership. The following year, he published the longer *Tractatus Apologeticus*, which expanded the same themes. Also appearing in 1617 was the first part of Fludd's major work, the *Utriusque cosmi . . . Historia* (The Technical, Physical and Metaphysical History of the Microcosm and Macrocosm), which was dedicated to God and James I. The initial issue was only the First Treatise of the First Book (*The History of the Macrocosm*), but contained the dedication and title of the whole work. The remainder of the *History* was published from 1618 to 1621.[3]

Apparently some "envious persons" called James' attention to the fact that a book was dedicated to him by an author who had written an apology for the Rosicrucians that contained spurious philosophy, innovation in religion, and heresy. His accusers claimed that Fludd had collaborated with the Rosicrucians in writing his *History*. But Fludd replied that he had composed his *Macrocosm* four or five years before he heard of the Brotherhood, (i.e., c. 1610–11).[4] He was particularly incensed by the fact that someone of rank who should have known better agreed with his accusers.

> . . . (saye they) we hould it unpossible that he should intreat in wright of such mystical learning (as they please to tearme it) and to be conversant in so many sciences, except other notable heads

were joyned with his in such a tedious business. Amongst the rest, this very concept was put into the head of an honorable personage and peere of this realme by on who seemeth a great schollar and Doctor.[5]

These, of course, were not charges that were to be taken lightly; had they been sustained, Fludd could have been in great difficulty, with severe consequences that could have included imprisonment or death. To answer the charges against him, James ordered Fludd to appear before him and explain in person what the work was about that carried the dedication to the monarch. A number of years later, Fludd gave this account of the audience:

> . . . King James of everlasting memorie for his Justice, Pietie, and great Learning, was by some Envious persons moved against mee, . . . but when I came onto him, and hee in his wisdome had examined the Truth and circumstance of every point, touching that scandalous report, which irregularly and untruly was related to mee, hee found me so cleare in my answer, and I him so regally learned and gracious in himself, and so excellent and subtill in his inquisitive objections, as well touching other points as this, that instead of a checke (I thank my God) I had much grace and honor from him, and received from that time forward many gracious favors of him: and I found him my just and Kingly Patron all the dayes of his life.[6]

Subsequent to his meeting with the King, Fludd wrote the "Declaratio Brevis," dedicated to his new patron.[7] James apparently thought it was a good idea for Fludd to write a short explanation about his works that would include the points covered during the audience. Thus he says at the beginning, " . . . I have composed, on your gracious suggestion, a Declaration, not an Apology . . . "[8]

The "Declaratio" is a Latin manuscript now in the Royal Manuscripts collection of the British Library (Royal MS 12C11). It contains ten small leaves and was written by the same amanuensis who wrote the first forty-three folios of a follow-up document to the "Declaratio," the "Philosophical Key" (discussed below). Another copy was destroyed in a fire. Although it was known to scholars for a number of years, it was not edited and published until 1978.

Presumably following the line of argumentation he took in defense of his work before the King and his attackers, Fludd opens the "Declaratio" with the reasonable statement that:

> . . . For who, understanding the depth of your Majesty's judgment, would not be disinclined to dedicate his books and writing to him [if they contained errors] or [subject them] to his scrutiny, by whose eyes, as if they were the eyes of a lynx, he examines every corner of a writer or speaker, and immediately discovers those errors in the works of others which are not . . . perceived by the . . . authors themselves.[9]

Indeed, considering Fludd's social and professional status, he certainly would not have presumed to dedicate his labor of twenty years to the King if there had been any thought at all that James would find his philosophy objectionable.

Fludd then proceeds to defend the *Tractatus Apologeticus* against the charges of advocating "religious innovation" and suspicion of heresy. First of all, before God and the King, he swears that he has unswervingly adhered to reformed Christianity as practiced in England since his infancy, and that he personally has remained completely chaste to that day. Indeed, the Brothers of the Rosy Cross are recognized by their German neighbors to be firm Calvinists, and the Rosicrucians themselves confirm this in the *Confessio*.[10]

The prime purpose for the publication of his apology, Fludd insists, "pertains to the impediment of the Arts, which are in a state of decline, and the method of reviving them; and then afterwards considers the wondrous qualities of Art and Nature with Philosophical arguments and by frequently using the affirmation of the ancients."[11] What makes this statement of particular interest is that it gives a succinct description of the basis for all of the many pages of folio volumes he wrote during his lifetime, which were accompanied by the most elaborate set of copperplate illustrations of any philosophical works ever published.

The Rosicrucians, Fludd says, earned his respect for two reasons. They profess to know the "true basis of natural philosophy, commonly unknown to this day," and they say they know the "profound secret of medicine," that is, they know how healing really takes place through spiritual means, by natural magic as it were. These two "lamps" for his mind, which are interrelated, were stated by the Brothers in two propositions:

> I. The true philosophy, commonly thought of as new, which destroys the old, is the head, the sume, the foundation, and the embracer of all Disciplines, Sciences and Arts. This true philosophy will contain much of Theology and Medicine but little of Jurisprudence; it will diligently investigate heaven and earth, and will sufficiently, *by its images* [emphasis mine] explore, examine, and depict Man, who is unique.
>
> II. We are able to show certain truths . . . by which . . . various illnesses can be cured. These truths are not to be divulged in a common manner . . . but in a new way . . . which is most certain and infallible.[12]

It is difficult to imagine two statements that would be more in accord with the entire thrust of Fludd's lifelong learning and work. The idea of a one true philosophy as the basis for investigating all man's endeavors and man's uniqueness itself is a precise statement about Fludd's Renaissance Christian Neoplatonism, as discussed in later chapters. He was also clearly intrigued by the esoteric healing methods which were directly linked with the first proposition.

Fludd himself makes the link by assuring the King that if he read about the lives of the "most distinguished" philosophers and physicians, he would find that they shared a common experience: to learn the mysteries of the divine philosophy, they traveled widely to visit wise men in Ethiopia and high priests in Egypt, read the inscriptions on the Pyramids, and studied natural magic with magi of Babylonia, Persia and India. Some of them even became worthy of the title of divine, even in Christian eyes, because they learned of, and acknowledged, the Trinity. Plato and Hermes Trismegistus are particularly distinguished among the philosophers for this honor, but Fludd also names Pythagoras, Thales of Melissus, Aristotle, Anaxagoras, Empedocles, Orpheus and Apollonius of Tyana as worthy of note. As for the physicians of note, Fludd designates Apollo, Aesculapius, Chiron the Centaur, Hippocrates of Cos, Chrysippus of Sicily, Aristratos of Macedonia, Euperices of Trinacria, Herosilus of Rhodes and Galen. These are basically a Renaissance Neoplatonist roll-call of important figures which stems in large part from the fifteenth-century revival of Platonism in Florence under the hand of Marsilio Ficino and Pico della Mirandola, which is discussed below.[13]

Finally, Fludd argues many wondrous secrets of nature remain unknown.

> ... And since the honorable and lawful inquiry into the
> mysteries of nature has as yet been by no age ever been
> prohibited or interdicted for the Philosopher and Physician, I,
> who profess to be a Philosopher and Physician (although not the
> most distinguished) have decided to offer this brief Declaration
> of mine to Your Most Memorable Majesty (even if I had not
> been urged to do so by your gracious suggestion) so that I may
> give satisfaction to your Majesty concerning the reasoning of my
> previously published *Tractatus Apologeticus* ... [14]

Moving on to his *History of the Macrocosm*, Fludd says that he could write a very large volume in defense of each part of it, which no doubt he could have, but it "would be wearisome for your Majesty to read", which no doubt it would have. In his following explanation of why the book was dedicated to God and James I, Fludd gives us the interesting information that he had entrusted his precious manuscript to someone in England who had carried it to the publisher, de Bry, in Germany, where an argument ensued between the carrier and the publisher over the dedication. The former wanted it dedicated to his own prince, the Landgrave of Hesse, which gives rise to the possibility that the carrier was Michael Maier. This is discussed further in Chapter IX. Fludd stepped in and insisted on a dedication to God and his own King.[15]

Fludd wants to show that he was not indiscreet in making it by inserting testimonials from "the Learned of every profession, by Papist and Lutheran as well as Calvinist," and he considered "that no one in the entire world was more worthy of the honor and dedication of these labors of mine, which have gained so much approval from the learned, than your Majesty ... "[16]

In fact, before the printer would publish Fludd's book, he sent it out for scholarly review by "many men of letters, both Papist and Lutheran as well as Calvinist", from whom came approval, even from the Jesuits (who only objected to the section on geomancy). Fludd therewith appends laudatory letters from Germany, Austria and France that give high praise to his work; the most distinguished of the writers was Gregor Horst (1578–1636), the "German Aesclepius."[17] Following Horst's letter, Fludd concludes by saying, as he did on a number of other occasions while defending his works, that unlike previous writers, who had disguised their true meaning in cryptic ways so as not to reveal divine truths to the unworthy, he has done the opposite:

> Certainly from the above letter from this most celebrated man it is sufficiently obvious that my opinions are not new, but rather are the most evident explications and most clear demonstrations of the secrets of nature which have been concealed or hidden by the ancient philosophers under the guise of allegorical riddles and enigmas.[18]

The "Declaratio" then closes with a florid entreaty to the King, "Astraeas's descendant," to judge his work with approval. Perhaps Fludd learned courtly manners from his father, who was so well respected at the court of Elizabeth.

However that may be, Fludd's supplications were obviously successful, since he says, as quoted earlier, that the King gave him "much grace and honor" at the time of the interview, and "received from that time forward many gracious favors from him: and I found him my just and kingly Patron all the days of his life." This does indeed appear to be the case. As discussed further below and in Chapter II, we have clear evidence of such favor from the documents about Fludd's application for a patent to make steel.

Since the "Declaratio Brevis" is in the Royal Manuscript collection, it is reasonable to assume that Fludd submitted the manuscript to the court, and that it may have been read by, or to, the King. Once he obtained the King's approval, however, there were still those at court to deal with who had denounced his works, and all others of their kind. Thus he dictated another manuscript which was intended for publication, "A Philosophical Key."[19]

The "Declaratio" must have been written about late 1618 or in 1619, because the date of the latest letter included is August, 1618, and the text only speaks of the *History of the Macrocosm*, which was published in two parts in 1617 and 1618. The "Philosophical Key" mentions both the *Macrocosm* and *Microcosm* books, the latter of which was published in five parts from 1619 to 1626, and it is therefore presumed to be the later manuscript. The first forty-three folios of the "Key" are written in the very legible script of the same amanuensis who wrote the "Declaratio."

The intent of the "Philosophical Key" is made clear on the title page and in the dedication to James I. It is an "ocular demonstration" which will open and decipher the hidden mysteries of nature by means of a chemical experiment and intellectual speculation about its results. It is another "Declaratio", this time in English, written to the "distrustful and suspicious" in order to show that his works are entirely his own, and to clear up spurious reports by ignorant, envious persons. In the dedication, he says to the King that the work contains the key to "unlock and open the meaninge" of the "Macrocosmical and Microcosmical Philosophy" which had been dedicated to his Majesty.[20]

As indicated in later chapters on Fludd's philosophy and the Rosicrucians, the "Philosophical Key" is quite helpful for understanding his work and personal relationships. It contains the previously mentioned title page, and a dedication to James, draped in courtly mannerisms, which are followed by three sections: "A Preface to the unpartial Reader"; "A Nosce te ipsum to the malitious Detractor, or the Calumniator's Vision"; and "Of the Excellency of Wheate"; a short Epilogue closes the work. The "Calumniator's Vision" is an abbreviated description of Creation and the development of the microcosm and macrocosm woven through the device of alchemical allegory; and the last and longest part is a step-by-step description and interpretation of Fludd's alchemical experiment on wheat, which he considered the basis of his entire philosophical edifice. Since there are notes to the publisher on where to put illustrations, it was obviously meant for publication. The experiment on wheat appeared later in Fludd's *Anatomiae Amphitheatrum* (1623), but the "Philosophical Key" did not make it into print until 1979.

In the Preface to the "Philosophical Key" we find more revealing information about Fludd's dealing with detractors at court. From his description of the encounter with these "calumniators," it appears that he confronted them during the same session or sessions at which the King was present, whom he satisfied with his replies to royal questions. Turning to the others, Fludd says that his greatest accusers are those who had examined his Apology the least. Indeed, when he questioned them, some "ingeniously and voluntarily confessed" that they had not even read the book, but only heard of it. Others had read it,

> . . . but not understanding the tenth part thereof (as by just
> inquiry it hath been plainly discovered) suspect my religion;

alleging that the places of Scripture by me produced are falsely and erroneously cited and strangely wrested from their true sense, and . . . that my philosophy is both spurious and corrupt, and also that my Theosophy is indirect.[21]

Another learned gentleman present insisted that Fludd had absolutely endorsed the Rosicrucian propositions without reservation, but was soon put straight by specific citations from the Apology, and thereupon acknowledged his error. As for his religion, Fludd adamantly maintains his steadfast adherence to Reformed Christianity, "so happily celebrated heare in England"; and finds it absurd to claim that he is an innovator in religion just because he defends the Rosicrucians, none of whom he had ever met. After all, didn't Cicero advise us to love those things that are virtuous, even if we haven't seen them?[22]

Turning to a defence of his monumental *History of the Microcosm and Macrocosm*, Fludd gives us a synopsis of his structural methodology, which is helpful as an overview. First of all, he says that he cites both sacred and philosophical authority to back up his statements and therefore they could not have been "newly by me invented." Secondly, he quotes "infallible axioms and perspicuous reasons out of ancient and true Philosophy" to underpin his arguments, which makes them neither irregular nor unreasonable. Thirdly, the foundations of his philosophy, he says, are in concert with that of Moses, "that is to say, Darkness, Water and Light." Lastly, he proves and makes plain his philosophical doctrine by "easy, familiar and ocular demonstrations, which is able to instruct the rude and rusticall, yea and very fooles themselves." Indeed, Fludd maintains that the ancient philosophers cloaked their doctrines in enigmas, elliptical references and ambiguous riddles in order to keep divine knowledge from the unworthy and ignorant; his philosophy, on the other hand, brings the mysteries of Nature from the shadows into the light.[23]

Woven between the same basic letters he appends to the "Key" to bolster his arguments about the acceptance of his writings by the learned community (which are the same as the ones in the "Declaratio"), Fludd tells of his disappointment at his unworthiness to be contacted by the Brothers and the Rosy Cross. He insists that he had written his *History* four or five years before he heard of them, and this would preclude his collusion with them in the writing of his great work, as had been alleged by the detractors. He goes on to say that his entire philosophy "hath flowed only from mine owne invention as Well in Practice as speculation."[24] His speculations were derived primarily from the experiment on wheat, which " . . . hath been the mayn practical perspicill or looking glass, in which I beheld the most asseured and continuated progress in Nature by which I composed the whole fabrick of my Philosophy . . . "[25] Furthermore, he finds that his "speculation which I have gathered out of this my practice in Alchemy" is not dissimilar to the tenets of the Rosicrucians, even though he knows them only through their books.[26]

Finally, the Preface to the "Philosophical Key" concludes with a statement of the "Courteous Reader" that the truth of his statements will be evident from the Reader's impartial inspection. On the other hand, he addresses a long, florid passage to the "Calumniator", beseeching him to know himself and his origins better, otherwise his charges against Fludd will only rebound back to their originator.

> If thou art ignorant what man is, and therfor pleadest for thy self, ignoramus, I know him not; I wil forthwith express and set before thyne eyes an object or looking glass or Nosce te ipsum, that from hence forth, al excuses layed a part, thou mayest noe more reply, Not guilty.[27]

Then follows the mythical allegory of creation with the development of the macrocosm and microcosm and the alchemical experiment with wheat.

The "Philosophical Key" was never published as written, and speculation on why this may have happened is covered in the chapter on the Rosicrucians. It probably remained in Fludd's possession, and at his death may have passed to Trinity College, Cambridge through his nephew, Lewin Fludd (d. 1678), who was an M.D. graduate of Trinity to whom Robert Fludd willed his library.[28] With the "Declaratio Brevis" and the "Philosophical Key," Fludd had answered his critics, and in the process, gained royal favor. The evidence, in addition to Fludd's own testimony quoted earlier, is clear.

As noted in an earlier chapter, Fludd was granted a patent to make steel in 1620. Since the timing of the patent application and the "Declaratio Brevis" and "Philosophical Key" correspond, it may be useful to look at them together. In May, 1618, the two holders of royal patents to make steel in England complained to the Privy Council that "one Robert Floude, doctor of physicke, hath entertayned a Frenchman to practice the same, hath sett up furnaces, and is now in worke [making steel], notwithstanding his majesty's inhibicion in the letter patentes."[29] The matter was referred to the Barons of the Exchequer to suppress the unauthorized steel making as they saw it.

It was shortly after this time, probably late 1618 or 1619, that the "Declaratio" was written, and probably 1619–20 when the follow-up "Key" was composed. As outlined in Chapter II, and above, Fludd had sometime earlier brought a French technician, John Rochier, to set up a large enough furnace to experiment with making steel. Two years after the initial complaint of the patentees, and clearly after his interview with the King and the writing of the defences of his work, Fludd petitioned James for a new patent. The King then orderd the Privy Council to take it up on 30 May 1620.

> Their Lordships did this day, according to his Majesty's gracious pleasure signifyed unto them, take consideration of a peticion presented to his Majestie by Robert Fludd, doctor of physick, wherein hee complaineth that havinge at his great expence and charge drawen over hither from forraine parts certaine persons expert and skillfull in makinge of steele, with purpose to imploy and sett them on worke here in that mystery, for the benefit and beoofe of the publicke.
>
> Forasmuch as it is conceaved that the makeinge of good and serviceable steele within the kingdome is a matter of good consequence, and worthie of all due encouragement; Their Lordships did entreate the Earl of Arundell, the Lord Digbie and Mr. Chancellor of the Exchecquer . . . to take notice of the foresaid peticion . . . [30]

With the King's backing and a presumably favorable report from the Barons of the Exchequer, at Hampton Court, on 27 September 1620, the Privy Council granted Fludd a patent to make steel. Present at the meeting were:

> The King's most excellent Majestie, Lord Archbishop of Canterburie, Lord Chancellor [Francis Bacon], Lord Stewarde, Lord Admirall, Lord Marquis Hamilton, Lord Chamberlin, Earl

of Arundell, Earl of Southampton, Earl of Kellie, Lord Viscount
Doncaster, Lord Digbie, Mr. Treasurer, Mr. Comptroller,
Mr. Secretarie Naunton, Mr. Secretarie Calvert, Mr. Chancellor
of the Exchecquer, Master of the Roles, Master of the Wardes.[31]

The council record noted that lately they had often debated the
making of steel in the kingdom, and because of the poor quality of
their product, had revoked the patents given to those who had
complained about Fludd's encroachment two years earlier.

"Robert Fludd, doctor of phisick," who had brought to England
John Rochier and other "operators" to make steel, presented the
Council with a certificate signed by many "cutlers, blacksmyths,
locksmyths and other artificers workinge in steele" that Rochier's
produce was "serviceable, good and sufficient." Furthermore,
Fludd offered to manufacture steel as good as any made elsewhere,
and sell it at a more reasonable price. It would be made using coal
alone, thereby saving precious wood, and restrictions on imports
would not be required. Last, but probably not least, the King
would get one-third of all profits. Thereupon the decision was
made:

> His Majestie and their Lordships, upon consideracion thereof,
> findinge it very requisite to have the makinge of good and
> serviceable steele settled within the kingdome, did well approve
> of the others made by Doctor Fludd herein, and think fitt that
> letters patents bee graunted unto him and the said John
> Rochier . . . [32]

On 27 November 1620, the council gave a warrant to the Solicitor
General to prepare the patent: " . . . this shal be to pray and
require you to prepare a graunt unto Doctor Fludd readie for his
Majesty's royall signature . . . "[33]

There is no question that Fludd's petition to James was
presented subsequent to their meeting where the King gave his
approval of Fludd's publications and that the monarch did act as a
patron in having the Privy Council consider the request, as Fludd
maintained. In light of this evidence, it cannot be maintained that
Fludd was an occultist outsider; he was in fact the recipient of
royal patronage following an examination of his philosophy. As
noted in the previous chapter, another interesting aspect of the
steel-making patent is that his experiments with chemistry resulted

in a very practical, useful, highly desirable product which would prove advantageous for both England and the crown. This in turn would have the effect of enhancing his reputation and prestige in the highest circles.

And in fact, royal patronage did not end with James I. For some reason that is not presently known, Charles I gave Fludd a valuable hereditary grant of the income from a manor in Suffolk in 1629:

> June 8 Westminster: Grant to Robert Fludd, Doctor of Physic, of a messuage and land in Kirton, Co. Suffolk, come to the Crown by being devised by Richard Smart to Anne Deletto, an alien, to the use of Rosamund Hewett to hold the same to such uses as by the will of Smart are limited.[34]

Whether this was a reward for services rendered or a kind of retainer fee is not clear, but it distinctly indicates royal pleasure with Fludd's work or services. That the grant came expressly from the wishes of the King is indicated by a note written at the bottom of the deed by Secretary Heath for Charles I's information:

> Whereas Richard Smart gent. did heretofore by his last will and testament devise to one Anne Delletto, to the use of Rosamund Hewett and her heires a messuage and divers lands in the countie of Suff; but by reason of the said Delletto is an alien borne the same are in strictures of lawe come to your Ma:tie. Now this conteyneth your Ma:ties estate and interest in the same messuage and lands to such uses as by the will is lymitted. Signified to bee your Ma:ts pleasure by Sir Sidney Mountagu. R. Heath.[35]

Although we do not presently know exactly why this valuable grant was made, or whether it was for a particular service or ongoing services, there can be no doubt whatsoever that Fludd continued to enjoy royal favor and patronage. It may have been that during the turbulent times of the Thirty Years' War on the Continent, Fludd's contacts there and occasional trips provided Charles with valuable informal information from someone he trusted. Further information about Fludd's relationship with Charles may prove difficult to come by, but would surely prove to be of great interest.

By the time of the granting of the patent to make steel in 1620,

Fludd was well established and secure in several respects. He had a prosperous medical practice; had published a major work in several volumes that gained praise from some colleagues and the patronage of the crown; and had produced high-grade steel from his chemical works which earned official praise and a royal patent. Fludd must have seen this success as a vindication of his beliefs; from some in England, Fludd's metaphysical speculations met with approval; from others, there was no doubt lack of interest or worse. But for the practical English, Fludd's tangible results in both medicine and chemistry were sufficient for him to enjoy a position of respect. It may also be worthwhile to note here that in addition to the metaphysics in his *History*, Fludd also had sections on the "practical" arts, such as surveying, use of perspective in art, and fortress-building. Thus while there were academic arguments over the metaphysics (primarily from some Continental scholars, as discussed in the following chapter) the fact is that, in his own time, they were not sufficient to tarnish his reputation in the highest circles; on the contrary, it was based on the general acceptance of his theories (or lack of serious objection to them), and the practical results of his activities. This then, was the background for the writing and publication of the rest of his books, and the academic polemics which followed in their wake. Without this background, too much emphasis in the past has been placed upon Fludd's metaphysics and the printed debates over them that followed, which has resulted in a distorted picture of Fludd's place in the early seventeenth century.

Chapter V

THE RENAISSANCE COMPLETED: PUBLICATIONS AND DEBATES

It was not for nought the wise man said: he that addeth unto
himself Science, contracteth unto himself much pain and
vexation, because that in science is much indignation.

Fludd, *Dr. Fludds Answer unto M. Foster*

For over twenty years, from 1617 until past his death in 1637, the
appearance of Fludd's publications provoked controversy and
debate. The nature of these arguments tells us much about both
view and may even point in that direction, but in the context of the
seventeenth century. We see those who completely agreed with
Fludd, as he shows from letters included in the "Declaratio
Brevis"; and those who, if they did not agree with him *in toto*,
certainly did in principle, such as his dedicatees and various
colleagues, which must include James I, the Bishop of Worcester,
the Bishop of Lincoln, the Archbishop of Canterbury, Sir William
Paddy and William Harvey, among others. Then we also see those
who attacked Fludd's works for various reasons, but from our
modern point of view, the attacks themselves, which may have
validity in saying that Fludd's theories do not represent the true
picture, would replace them with views that were not, in many
respects, valid either. Some of them contain elements of a modern
view and may even point in that direction, but in the context of the
times it is far too facile, and even incorrect, to say that it is a
debate between the ancients and the moderns, between the archaic
imagination and the now-accepted notions of quantitative science.
But it could be said that if Fludd's great philosophical edifice,
which was based on the Renaissance Christian Neoplatonist view
and elaborated in great detail with many marvelous illustrations,
would not hold up, more than just his philosophy was at stake: the
whole notion of the metaphysical foundations of the cosmos, the
collected wisdom of millennia, and, indeed, the source of our
universal truths hung in the balance. The Renaissance itself, as an
intellectual movement to recover lost ancient knowledge and make

use of it to unify and reform the world, was also on trial.

It is necessary to remember also in this context that Fludd's books were, in his own lifetime, considered to be serious, important works. From the high-ranking personages to whom they were dedicated, and from admirers on the Continent, he received praise and encouragement; but even the fact that Kepler, Mersenne and Gassendi on the Continent chose to criticize him at such length shows that the Fluddean corpus was not to be dismissed lightly, that indeed he was important enough to warrant engaging in lengthy debate in print by scholars who are now considered to be important figures in the emerging Scientific Revolution.

Fludd's publications and the controversies and debates which followed should help illustrate his position at the end of the Renaissance, and indeed at the end of two millennia of accumulated speculative thought. The publications themselves were a unique achievement. Although most of Fludd's constructions were readily available in whole or part from many ancient and Renaissance sources, they were elaborated, coordinated, synthesized and illustrated in greater detail than in any other set of speculative works before or since. They presented a summation and synthesis of the universe according to the Renaissance Christian Neoplatonist perspective, which was in some places interpreted by some of Fludd's own idiosyncratic inventions. Above all, the mass of detail about the upper, lower and intermediate realms was integrated into a comprehensive whole; in a word, he created a completely interwoven system that rivaled Aristotle's, which is exactly what he had in mind. In the quality and number of illustrations alone, perhaps only Diderot's *Encyclopedia* a century and half later superseded Fludd's books, and one still finds them appearing in a variety of modern publications, including the collected works of Shakespeare.

Fludd's first printed book was the brief Rosicrucian defense, the *Apologia Compendiaria* of 1616, which was followed up by the expansion of this outline the following year, the *Tractatus Apologeticus integritatem societas de Rosea Cruce defendens*. The controversies they caused for their author are discussed in the previous chapter and the chapter below on the Rosicrucian problem. Undoubtedly, his greatest publication was the uncompleted *Utriusque cosmi . . . historia*, which was issued in five parts by de Bry from 1617 to 1621 (see Bibliography). This *Technical Physical and Metaphysical History of the*

Macrocosm and Microcosm, which covers some seventeen hundred folio pages in all, was the result of over twenty years' work, stretching back to his student days at Oxford, when he carefully began to organize his broad-ranging Renaissance studies into treatises on metaphysics and the arts and sciences. By the time the various parts were pulled together as an integrated whole, he had created a comprehensive philosophical system based on a multitude of ancient and Renaissance sources. But it was not his Renaissance Christian Neoplatonist system which kept an interest in his books alive over the centuries; it was the fine copperplate illustrations, which are so abundant in the *Utriusque cosmi . . . historia*.[1]

The first part of the *History* was issued in 1617 under the title of the whole work, and comprised the first tractate of two on the *History of the Macrocosm*. Here Fludd begins with creation, which is in accord with both Genesis and alchemical principles, and then details how the macrocosm came to be structured from it: the harmonic relationships between the parts, and the three-part division of the macrocosmic heaven into the Empyrean, Ethereal and Elementary (Plate 17).

Tractate II of the *Macrocosm* bore the title *Tractatus Secundus. De Naturae Simia . . .* and contained chapters on "Universal" arithmetic; the Temple of Music, a device to show in one graphic construction all musical harmonies; Geometry; Optics; the Pictorial Arts, including the use of perspective; Motion; Time; Cosmography; Astrology; and Geomancy, or a method of divination using random dots that were originally drawn in the earth, thus "geomancy" (Plates 18 and 19).

In 1619, Johannes Kepler published his *Harmonices Mundi* in Linz, Austria, which contained his theories of the harmonics of the universe; it also had attached a long appendix refuting Fludd's harmonics in the *History of the Macrocosm*. Fludd replied to Kepler in his *Veritatis Proscenium* of 1621, in which he systematically rebutted Kepler's objections point by point. The following year, Kepler continued the polemic with his *Pro suo opere harmonices mundi apologia*. Here Kepler clearly draws the distinction between his own mathematical and Fludd's alchemical-Hermetic approach to knowledge of the physical world and its relation to the divine. The final round of the volley was Fludd's *Monochordum Mundi*, which was attached to his mystically-based work on anatomy, the *Anatomiae Amphitheatrum* of 1623. (The *Anatomiae Amphitheatrum*, 1623, and the *Philosophia sacra*, 1626, which dealt with meteorology and cosmology, were the last parts produced of the *Microcosm History*.)

On the surface, the polemical sparring between Fludd and Kepler could be viewed as an argument between a Hermetic-Cabalist-alchemist mystic and a proto-modern scientist who insisted on the primacy of exact mathematical proofs for his discoveries. On the other hand, it could also be considered as a difference of opinion between two followers of Plato who disagree about their versions of the harmonics of the Pythagorean-Neoplatonist structure of the universe. Neither version gives us a truly accurate picture of the basis for their disagreement; and it is exactly this basis that provides us a key to understanding the climate in which they wrote.

Kepler was in fact an ardent Pythagorean-Neoplatonist as well as a pioneer mathematician. But it is altogether noteworthy that his famous mathematically-discovered laws of planetary motion were stepping stones to serve his *a priori* conception of the universe as made up of the five regular Pythagorean polyhedra, which fit neatly one inside the other, with the orbits of each of the individual planets contained in its own regular polyhedron. Kepler was struck with this notion in an intuitive flash while teaching a mathematics class in Gratz, Austria, in the year 1595, when he was twenty-four. It came as an answer to a question, but one which had a specific frame of reference: why were there only six planets? The solution had to fall within Kepler's orientation as a Renaissance follower of Plato and Pythagoras. When he wrote to Galileo in 1597, for example, he said: "By the strength of your personal example you advise us, in a cleverly veiled manner, to go out of the way of general ignorance and [warn us against exposing ourselves to] the furious attacks of the scholarly crowd. (In this you are following the lead of Plato and Pythagoras, our true masters.)"[2]

In 1607, he wrote to James I (who had visited Tycho Brahe's observatory in Denmark) when he sent the King a copy of his new book, *A New Star in the Foot of Serpentarius* (which was dedicated to James): "To the King-Philosopher, the serving philosopher; to the Plato, the ruler of Britain . . ."[3] And of course his writings abound with references to Plato and Pythagoras, as well as to his favorite Neoplatonist, Proclus. Thus when the solution struck him, it appeared to be perfect; for the rest of his life he was utterly convinced of its truth, and set about proving it by every means available to his faculties, which included his considerable mathematical skill. For the next six months after his "discovery," the young *mathematicus* struggled to write a book that would incorporate all the proof he could bring to bear. The fruit of this effort was the *Mysterium Cosmographicum* of 1597.

The *Mysterium* established his thesis that Copernicus was right about the centrality of the sun, and that God created a perfect universe where the five symmetrical solids may be exactly placed between the six orbits of the planets. Also included are chapters on astrology, numerology, the geometrical symbolism of the zodiac and the Pythagorean harmony of the spheres, in which Kepler looked for correlations between the five perfect solids and the harmonic intervals of music. These chapters place Kepler squarely in the Renaissance Neoplatonist stream; but in the latter part of the book he turns from speculative thought to an examination of his theory by mathematical means, and can't quite make it work. The book was read with some interest by other Renaissance Platonists, including Fludd, but rejected by many, including Galileo.[4]

Already in the *Mysterium*, however, we can see the similarities and differences between Fludd and Kepler begin to emerge. They shared a Renaissance Neoplatonist view of the cosmos: that it was exactly proportional and hierarchical in structure according to Platonist-Pythagorean teaching, and infused with non-material forces; that there was a distinct relationship of the material to the non-material, and that the whole was knowable by the human mind. But major differences began to show themselves as well, when Kepler, who was committed to Copernicus' theory, insisted on proving his case mathematically, which he was convinced must be possible from the observed data. When he fails to do so in the *Mysterium*, he blames the faulty Copernican data. One can imagine this would reinforce the views of the non-Copernicans like Fludd, who would agree with a scheme of Pythagorean harmonies and other parts of Kepler's book but not the sun-centred universe or the planetary orbits fitting into the five regular solids.

From 1601 to 1612, Kepler held the post of imperial mathematician at the court of Rudolph II in Prague, where he was able to make use of Tycho Brahe's unique twenty-year compilation of new observations. During this time Fludd traveled on the Continent, returned to England to finish his medical degree, established a practice in London, and finished the manuscript of the *History of the Macrocosm*. In 1609, Kepler published the *Astronomia Nova*, which he dedicated to Emperor Rudolph, that made use of Tycho's data for the first time. It contained the pioneering first and second of his laws of planetary motion: that the orbits are elliptical with the sun at one focus, and that a line drawn from the sun to the planet sweeps out equal areas in equal times.

But Kepler's discoveries were difficult for his contemporaries to accept, including Galileo. Without Newton's physics and the law of gravity, the laws appeared to be mere mathematical constructions without any logical conceptual framework for them to fit in. Why the orbits should be an ellipse was a troublesome philosophical question that bothered Kepler himself, who wanted to get on with proving his pet theory about the five regular solids. Another problem was that few scholars at the time were capable of following Kepler's difficult mathematical journeys on the way to his discoveries, let alone to recognize their importance.

After the death of Rudolph II in 1612, Kepler took the post of provincial mathematician in Linz, where he remained for the next fourteen years. It was here that he worked on his *Harmonics of the World*, which was completed in 1618. But before printing was completed, Kepler saw the *History of the Macrocosm* (Frankfurt, 1617 and 1618) and wrote the appendix that took issue with Fludd's version of the musical construction of the universe. It is important to note that the *Utriusque cosmi . . . historia*, of which the *Macrocosm* forms the first part, was dedicated to God and James I, and that Fludd, as discussed earlier, personally appeared before the King to defend the book, and thereafter received royal patronage. Kepler's *Harmonices Mundi* was also dedicated to James, with Kepler saying that "he could find no more suitable patron for his work, which follows that of Pythagoras and Plato, than the great King who demonstrates his partiality to the philosophy of Plato through his own accomplishments."[5]

Kepler obviously set great store on James I as a patron of followers of the Renaissance Pythagorean-Platonic-Neoplatonist school of thought. He no doubt knew that James had visited Tycho Brahe at Uraniborg years earlier, and, as mentioned above, had dedicated his earlier book of 1607 (which had made use of Tycho's data) to the King. With the *Harmonices Mundi*, Kepler was presenting his greatest achievement to a foreign monarch, the same one to whom Fludd had dedicated his own *magnum opus* (a fact that cannot have escaped Kepler's attention). In the books of both Kepler and Fludd could be found a comprehensive, systematic ordering of the cosmos according to their interpretation of the principles of Pythagorean, Platonic and Neoplatonic thought. There were clear distinctions between their assumptions, methods and results, but both were just as clearly a part of the broad Renaissance Christian Neoplatonist tradition, just as they both appealed to James I to be their patron.

Although Kepler is hailed as one of the key figures of the emerging Scientific Revolution for his discovery of the now-famous three laws of planetary motion, and his insistence on adhering to the dictates of the mathematical consequences of the observed data, he and Fludd occupied very similar positions in the intellectual world of the early seventeenth century. In the *Harmonices Mundi*, Kepler promulgated his third law (that the squares of the periods of revolution of any two planets are equal to the cubes of their mean distances from the sun), but once again, it was only a tool to be used to push forward his proof of the orbits inhabiting the five symmetrical Pythagorean polyhedra. The first law is only given slight mention, and the second law none. His physical laws were right, and marked a significant pioneering effort in the discovery of natural laws by mathematical analysis of observed data; but his quest was for proof of his Renaissance Neoplatonist concept of the universe, for which he sought the approval and patronage of James I. Fludd's quest was exactly the same.

Both men still inhabited a world where physics, ancient metaphysics and religion had not yet become separated. The physical world could only make complete sense within the framework of a divinely ordered universe. Fludd's steel-making was akin to Kepler's laws. In the course of alchemical experiments which were tied to Paracelsian-Neoplatonist metaphysics, Fludd discovered a high-grade steel process, for which he received a royal patent under the patronage of James I. Unfortunately for him, this fact fell out of sight until recently, and so Fludd became better known as the occultist Rosicrucian who argued on the wrong side of the mathematically-proven Copernicanism of Kepler. But Kepler's laws could only be vindicated by the genius of Newton much later in the century, whereas in James I's time the contest was much more evenly matched.

There have been some illuminating examinations of the differences between Fludd and Kepler.[6] In a 1967 article, Peter J. Ammann wrote an excellent summary of Fludd's musical theory and its relationship to his broader philosophical scheme, and elaborated the illustration through the polemics with Kepler.[7] Fludd's original notion of the celestial monochord divided the regions between God and earth into two octaves, with the sun exactly in the center; the lowest note, at the earth, corresponded with the most material, and the highest with the most spiritual, exactly in proportion to the scheme of the opposite intersecting pyramids. From top to bottom was further divided into the three equal realms, or heavens. Kepler objected in the *Harmonices* appendix that the pyramids compared light and matter, which cannot be done, and from this faulty concept Fludd had fashioned a fictitious world of his own. On the other hand, Kepler said that he was only looking for harmonies that can be determined by the planetary movements. He went on to say that "it is obvious that he [Fludd] derives his main pleasure from unintelligible charades about the real world whereas my purpose is, on the contrary, to draw the obscure facts of nature into the bright light of knowledge. His method is the business of alchemists, hermetists and Paracelsians; mine is the task of the mathematician."[8]

Fludd also maintained in many places in his writings that he was the first to make the enigmas and riddles of the alchemists, hermetists, etc. into a clear understanding of the workings of nature. But Kepler, Fludd responded in the *Veritatis proscenium*, "is concerned with the external movements of things, but I with the internal and essential processes of Nature." And furthermore, "the ordinary mathematicians deal with the shadows of quantities, the chemists and hermetists, however, grasp the true essence of natural things."[9] From a strict Platonist point of view, Fludd would be right: the external, visible world measured by mathematics is a mere shadowy reflection of the real, and vastly more important world. In Fludd's view, Kepler is dealing only with the material world, and ignoring the spiritual one.

Kepler's response of 1622, *Pro suo opere harmonices mundi apologia*, reiterated the basic differences between the two approaches and their resulting systems. Fludd's second and last reply in the matter was the *Monochordum mundi* of 1623 (Plate 7), in which he revised his original harmonic scheme (Plate 3) to make it more comprehensive and to respond to Kepler's criticism. This Universal Monochord, which bears the title, "God is all there is; from Him all things proceed and to Him all things must return," attempts to coordinate Neoplantonic hierarchical emanations from the Godhead to Earth with musical harmonies, a Cabalistic interpretation of the Divine Name, and a three-part division of the Tabernacle of Moses. The entire scheme was a comprehensive synthesis of Fludd's Renaissance Christian Neoplatonism complete with the Cabala. Although the concept of universal harmonies was quite old and available to Fludd from a variety of Renaissance and ancient sources (particularly Plato, Proclus, Marsilio Ficino, Agrippa von Nettesheim and Francesco Giorgi), his particular coordinated synthesis was unique.

A famous examination of the Fludd-Kepler debate was undertaken by the physicist Wolfgang Pauli (a 1945 Nobel Laureate) in a 1952 essay, "The Influences of Archetypal Ideas on the Scientific Theories of Kepler."[10] It was published in the same volume as an essay on "Synchronicity" by Carl Jung, who was an associate of Pauli's. His important perspective on the debate was that "as ordering operators and image-formers in this world of symbolic images, the archetypes thus function as the sought-for bridge between the sense perceptions and the ideas and are, accordingly, a necessary pre-supposition even for evolving a scientific theory of nature."[11]

From a careful elucidation of textual excerpts, Pauli shows that for Kepler, his formulation of a natural law was preceded by his symbolical conceptions, and that even his astrological and meteorological beliefs "were integrated with scientific-causal thinking."[12] He then contrasts Kepler's approach with that of Fludd, which he characterizes as an "archaistic-magical description of nature culminating in a mystery of transmutation."[13] From a like examination of Fludd's texts, Pauli concludes that "Fludd's symbolical *picturae* and Kepler's geometrical diagrams present an irreconcilable contradiction," which is generalized into two types of mind that may be found throughout history.[14] The first type is concerned only with the quantitative relations of the parts, while the second deals with the qualitative indivisibility of the whole (the "thinking" type vs. the "feeling" or "intuitive" type). The debate then, illustrates that the "quatenary" attitude of Fludd corresponds, in contrast to Kepler's "trinitarian" attitude, from a psychological point of view, to a greater *completeness of experience*. Whereas Kepler conceives of the soul almost as a mathematically describable system of resonators, it has always been the symbolical image that has tried to express, in addition, the immeasurable side of experience, which also includes the imponderables of the emotions and emotional evaluations. Even though at the cost of consciousness of the quantitative side of nature and its laws, Fludd's "hieroglyphic" figures do try to preserve a *unity* of inner experience of the "observer" and the external processes of nature, and thus a *wholeness* in its contemplation – a wholeness formerly contained in the idea of the analogy between microcosm and macrocosm but apparently already lacking in Kepler and lost in the world view of classical natural science.[15]

This radical shift from the qualitative to the quantitative point of view that marks the beginning of the Scientific Revolution and the end of the Renaissance has the possibility of finally being reconciled in modern, post-quantum physics as well as in psychology, in Pauli's judgment. The only acceptable outlook should be the one that "recognizes *both* sides of reality – the quantitative and the qualitative, the physical and the psychical – as compatible with each other, and can embrace them simultaneously."[16]

Deeper insight into Pauli's analysis and the Fludd-Kepler debate was provided by Robert S. Westman in a valuable essay, "Nature, art and psyche: Jung, Pauli, and the Kepler-Fludd polemic."[17] Westman maintains that Fludd's *picturae* are more than illustrations of the complex written texts: they are "ways of *knowing*, *demonstrating*, and *remembering*," which aid the beholder to direct the self back to inner unity with the Creator.[18] The common denominator that bridges the theoretical and practical in Fludd's iconography is the human form, which incorporates both the principles of geometry and all-important symmetry. A convincing argument is made that Albrecht Dürer provided Fludd with the primary source for this scheme: he specifically refers to Dürer's *Four Books on Human Proportions*, and several illustrations are taken directly from this work, the most important one being a key depiction of the harmonic proportions of man, the microcosm.[19]

Dürer had divided man into proportional parts as an instructional device on the drawing of the human form, but Fludd expanded the notion to be the centerpiece of the harmonic microcosm-macrocosm relationship.[20]

Fludd's central theme may be viewed as part of a large Renaissance body of publications of commentaries on Genesis, and that was combined with his strong belief that it was only through images that a complete grasp of the occult, mysterious, essential truth of the Word could be effected. As Fludd understood Genesis and played it out in his illustrations, light emanated from darkness, and these opposites were mediated by the Spirit to form the Trinity, out of which in turn the four elements developed and stabilized in a concentric configuration. The Creation process continued on successive days to fill out the three regions and all things in them in harmonic proportion. This pattern then permeates all the created world, as may be demonstrated by alchemical processes, the weather-glass, and even the functioning of the human body. The harmonics derived from Creation proceeded in the order of original unity, duality, trinity and quatenary, and therefore Fludd's harmonic scheme only included the Pythagorean octave (2:1), fifth (3:2), and fourth (4:3), which he often depicts as the pyramid [21]

In true Neoplatonist fashion, the goal is to ascend from the physical world, through contemplation, to the highest, to climb from the quadripart world to that of unity in God. The inner vision can be directed toward that goal through external images to reverse, as it were, the process of Creation. Thus the illustrations are far more important than mere aids to understanding the text; they are a vital medium that the imaginative Soul uses to connect the Intellect with the Sensitive Soul, in Fluddean Neoplatonic terms.[22] The whole structure is a completely consistent scheme that combines Neoplatonic Hermeticism, Paracelsian alchemy, Genesis and the Cabala into his *Mosaicall Philosophy*.

Kepler had his own archetypal images, as discussed above, as the Creator's grand plan, which centered around the five regular polyhedra; but his approach was really from the other end, the precise mathematical geometry derived from observations of planetary motion, and thus he found Fludd's image-making, stemming from his interpretation of Creation, not to be helpful in elucidating these "dense mysteries."[23] More than that, however, the impact of Fludd's ideas was sufficient to provoke an extended debate with Kepler, as well as with Mersenne and Gassendi, which is treated below. But rather than viewing Fludd and Kepler as opposites, they may be considered, in Pauli's terms based on Jungian psychology, as "complementary aspects of the same reality,"[24] and Westman has shown convincingly that the "true theme" of the Jung and Pauli essays that appeared together is a formulation of the resolution of seeming opposition between the quantitative/thinking view and the qualitative/intuitive one, as represented by Kepler and Fludd.[25]

Further insight into the essays comes from the fascinating discovery that it was the interpretation of Pauli's dreams that Jung used to formulate his idea of the mandala as the symbol of self.[26] The dreams appear in Jung's *Psychology and Alchemy*, in which he includes many illustrations from Renaissance alchemical texts to show that alchemical symbolism reflects psychic states described in the dreams. Taking pride of place at the head of the chapter entitled "The Initial Dreams," now known to be Pauli's, is one of Fludd's best known illustrations, "The Mirror of the Whole of Nature and the Image of Art," which Jung captioned "The *anima mundi*, guide of mankind, herself guided by God"[27] (See Plate 9). Further along in the book, Jung states that "the *anima mundi* [here the Fludd illustration is cited] coincides with that of the collective unconscious whose centre is the self."[28] Westman concludes that "Pauli must have believed that Fludd's pictures represented symbols of the collective unconscious and the self, and that by studying Fludd he was gaining access to the *Fluddean part of himself*," since he was experiencing this same conflict of visualization in quantum mechanics.[29] Modern representatives of this conflict may be seen in Heisenberg's pure mathematical formulations versus the imaging of Schrödinger or Bohr.[30]

An interesting footnote to the Fludd-Kepler debate is provided by some little-known facts. When, in 1626, Kepler left Linz, which had been under siege by rebellious Lutheran peasants, he went to Ulm with his manuscript of the Rudolphine Tables to find a printer. While there, he stayed in a house provided by a friend,

Dr Gregor Horst, the widely-known and respected Ulm town physician who had praised Fludd's *History* so highly in a letter of 1618 when he was Professor of Medicine at the University of Giessen and chief physician to Ludwig of Hesse.[31] Horst's letter was one of the most prominent ones Fludd included in the "Declaratio Brevis" and the "Philosophical Key."[32] Another is the fact that in 1620, Sir Henry Wotten (1568–1639), an English ambassador, witnessed some experiments conducted by Kepler in Linz, and sent an account of them to his close friend, Sir Francis Bacon, who was working on the *New Organon*. When Bacon's book was finished, he sent three copies to Wotten, who promised to send one to Kepler. During his visit with Kepler in 1620, Wotten invited Kepler to come to England, but Kepler decided against leaving his homeland.[33] One can only speculate about the England of James I's time with a Bacon–Kepler collaboration. Perhaps there would have been more sparring about Fludd's religious-hermetic metaphysics, but certainly not about his practical achievements in medicine and chemistry, i.e., steel-making.

In 1623, the year of Fludd's final reply to Kepler, a small tract appeared in London by a little-known figure named Patrick Scot, Esq., which bore the title *The Tillage of Light*.[34] Scot's thesis was that the descriptions of alchemical processes were but allegories of the steps involved in acquiring wisdom and bringing it to perfection. Fludd was moved to reply by penning a ten-thousand-word response, "Truth's Golden Harrow," in which he maintained that the philosopher's stone has a real existence, and that physical transmutations are indeed possible, but that the true philosopher is not concerned with such vulgar matters, only with the greater work, the transmutation of the soul. "Truth's Golden Harrow" was never published, but it contains a characteristic digest of Fludd's belief in the corresponding macrocosm-microcosm relationship of alchemy, which is a consequence of Creation as described by Moses in Genesis, and by Plato, Hermes, the Cabalists and others.[35]

Just as Fludd had concluded his polemic with Kepler in 1623 and replied to Scot's *Tillage of Light*, that same year a new controversy opened with the publication of another commentary on Genesis, *Quaestiones celeberrimae in Genesim*, by Marin Mersenne (1588–1648), the French mechanist. Mersenne, a friar in the Order of Minimes who resided in Paris, is most noted for the wide correspondence he maintained with many leading scientific figures of the day and his vigorous advocacy of a mechanistic world-view. His approach to Genesis was from an orthodox point of view which drew on the Fathers and traditional commentary, and he denounced those who would connect it to alchemy and the Hermetic texts. Indeed, he was particularly harsh in his dislike of the central tenets of Renaissance Hermetic thought and their adherents, and specifically singled out Ficino, Pico, Agrippa, Trithemius, Giorgi and others who are central figures in Fludd's philosophy, for condemnation.[36] Mersenne's attack against Fludd was specially vitriolic; at one point he even called Fludd an "evil magician, a doctor and propagator of foul and horrendous magic, a heretical magician," and in other places excoriated him for defending the Rosicrucians.[37] As an Oxford-educated English esquire descended from gentry on both sides Fludd was shocked by such tactless, vulgar vehemence from a member of the learned community. His counter-attack was printed in 1629, the *Sophie cum moria certamen*, in which he restates and defends his basic themes, including the use of chiromancy, a practice Mersenne had taken special pains to denounce. Fludd also complained that Mersenne had singled out passages from the *History* for criticism without saying who wrote them or where they came from, and that Mersenne had not grasped the fact that Fludd had clearly distinguished evil magic from the true and good one; had he done so, no doubt Mersenne would not have accused him of being a "cacomagus." The conclusion was aptly put by Allen Debus: "Convinced of the truth of his own views and appalled by the acidity of Mersenne's attack, Fludd could only conclude that it was the nature of his opponent to be violent, indeed, insane."[38]

Bound at the end of the *Sophie cum moria certamen* was another short tract of fifty-three pages entitled *Summum Bonum*, by a Joachim Frizius.[39] It is also a reply to the "calumnies" of Mersenne that seeks to defend magic, Cabala, alchemy and the Fraternity of the Rosy Cross as participating in the "highest good." The title page carries a large emblematic engraving of a rose growing on a cross stem, and has the motto DAT ROSA ME LAPIBUS arched across the top. Since it was bound with the *Sophie* and uses many typical Biblical, Hermetic and other Renaissance Neoplatonist sources, Frizius has often been assumed to be Fludd writing under a pseudonym; but he explicitly denied being the author, and said that it was the work of a good friend, which he evidently sent to the printer, perhaps after some editing. He anticipated the confusion by asking the printer to bring it out in a separate edition, octavo, instead of as a folio addendum to his own book; but the publisher, in a note to the reader, explained that they wanted to include it with Fludd's reply to Mersenne.[40]

In the midst of his replies to Kepler and Mersenne, Fludd also published the last two parts of the *History of the Macrocosm*, the *Anatomiae Amphitheatrum* and *Philosophia Sacra* in 1623 and 1626, respectively, as mentioned previously. Mersenne apparently felt the need for allies in his scholarly war with Fludd, so in 1628 he asked his friend Pierre Gassendi (1592–1655) to look at Fludd's books, along with that of his colleague, William Harvey, and prepare an analysis. The result was an *Examen philosophiae Roberti Fluddi* of 1630, which is usually known by its original title *Epistolica exercitatio*.[41] After summarizing Fludd's philosophical scheme, Gassendi turned to a critique of *Sophie cum moria certamen* and *Summum Bonum*, and at the end appended a letter from Francisco Lanovius, another Parisian scholar, who believed that tying alchemy to Scripture was blasphemy and sacrilege. Gassendi's conclusions were in accordance with those of his fellow mechanist Mersenne, but in more scholarly and respectful terms.

An odd aspect of Gassendi's *examen* shows the nature of argument in the early seventeenth century, when the Hermetic-alchemical camp as well as the old Aristotelian were under siege from the mechanists; but all the lines of demarcation were not clear, and as the Renaissance waned in this period curious mixtures are found. Gassendi rejected Harvey's theory of the circulation of the blood, but did so in the context of criticizing Fludd's theory of the spirit of the blood and its relationship to the microcosm-macrocosm relationship in the *Sophie*. Although Gassendi admired Harvey's experimental methods, he was convinced that the blood did not circulate and took the opportunity to refute the view while reviewing Fludd's work.[42] In an ironic twist, Fludd, about the time Gassendi had prepared his book, had written a work on the pulse in which he was the first to agree with Harvey's theory in print, and that it was in consonance with his own philosophy, which

exactly seems to confirm the feeling and opinion of the learned William Harvey, a most skillful doctor of medicine, most clear in the art of anatomy, and yet highly versed in the mysterious profundities of philosophy, a man who is my esteemed compatriot and the most faithful of the College: about which he instructed the world advisedly and prudently in his littel book *Exercitatio de cordis sanguinis in animalibus motu*, and he declared remarkably well as with demonstrations for the eye that the motion of the blood is circular.[43]

Thus we have the mechanist Gassendi arguing on the wrong side of blood circulation, and Fludd, the mystical alchemist, who had witnessed many of Harvey's anatomical experiments, arguing on the correct side.

Another interesting aspect of this intellectual sparring was Mersenne's discovery of Meric Casaubon's critical study of the Hermetic texts, which he concluded were not of ancient origin, but written in early Christian times.[44] Mersenne immediately passed on the information to Gassendi, while the latter was preparing his examination of Fludd's philosophy.[45] Once Casaubon's exposure of the *Hermetica* as early Christian forgeries was in the hands of mechanists and other foes of the Renaissance Hermetic adherents, the life of that school as a vital force was severely threatened. As Francis Yates put it,

> It shattered at one blow the build-up of Renaissance
> Neoplatonism with its basis in the *prisci theologi*, of whom Hermes
> Trismegistus was the chief. It shattered the whole position of the
> Renaissance Magus and Renaissance magic, with its Hermetic-
> Cabalist foundation . . . It shattered, too, the basis for all
> attempts to build a natural theology on Hermeticism . . . [46]

But in another way, Casaubon's critical study gave an additional irony to the story: his work containing the dating of the *Hermetica* was written at the encouragement of James I, and published in London in 1614 with a dedication to the King, some four years or so before Fludd made his defense before James of his Hermetic-Mosaic-Cabalistic-alchemical work and won royal favor![47]

Fludd did not waver in his conviction about the truth of his philosophy, and so, after the publication of Gassendi's *Examen* in 1630, he set to work on a final reply to Mersenne, Gassendi and Lanovius, which was completed by the following year. It appeared in 1633 under the title *Clavis philosophiae et alchemiae Fluddannae*, in which he systematically dissects the attacks of all three opponents and answers each part in turn. The section dealing with Gassendi includes a further defence of Harvey's theory of the circulation of the blood, backed by experimental evidence. The final part of the eighty-seven-page folio book is a reiteration and reaffirmation of the mystical-divine-alchemical origins of the world, and how it operates through the penetrating spirit; and the nature of true alchemy in transforming the human soul that was also the goal of the Rosicrucians, who were persecuted "because wisdom is, according to her name, to the unlearned unpleasant."[48]

Neither Mersenne, Gassendi nor Lanovius published anything more against Fludd, but a practically unknown response to the *Clavis* was written by a Jean Durelle, which was published in Paris in 1636. Mersenne tried to get the book republished in England, but without success due to lack of interest.[49] From Fludd's side, however, he had some last words to add to the debate. In the midst of preparing the *Clavis*, a new controversy popped up, this time from England. Parson William Foster (1591–1643), of Hedgly in Buckinghamshire, in 1631 published a tract against Fludd's advocacy of the weapon-salve cure as diabolical in *Hoplocrisma-spongus: or, A Sponge to Wipe Away the Weapon-Salve*. The weapon-salve was a preparation applied to the weapon causing a wound, which was believed to aid healing by sympathetic natural magic. Foster's complaint was not that the cure didn't work, but that it worked by using diabolical forces and was therefore evil, black magic. To bolster his attack, the parson used statements from Mersenne and Lanovious about Fludd being an evil magician.[50]

Since Foster's book was in English, Fludd replied in kind with the sixty-eight-page *Doctor Fludds Answer unto M. Foster or The Squeesing of Parson Fosters Sponge* in 1631.[51] In this small work Fludd not only rebuked Foster in the strongest terms and clearly showed the weakness of his arguments, but also shed light on the Mersenne–Gassendi polemic, since they were pulled into the argument. As for Mersenne's immoderate accusations, Fludd cites Gassendi's own rebuke:

And although (my Mersennus) the zeale wherewith you are moved against Fludd is to bee commended, neverthelesse you cannot bee ignorant, how grievous and intolerable a thing it is unto any man that liveth in the Christian world, to be called a witch, or evill Magitian, a Hereticke-Magitian, or a teacher, or divulger of foule and horrible Magicke: and that such a teacher is not to bee suffered unpunished, also to provoke the King or Prince to punish him, and besides all this to threaten him; saying that cause hee should bee drowned or drenched in the eternall Lake and so forth . . . I will not that any man should bee patient in the suspition of Heresie; much lesse to be accused or suspected of Atheisme or naughty magicke.[52]

Warming to his subject, Fludd says that in Foster's work he smells a rat in being accused of evil magic merely because of Mersenne's unfounded charges:

And who is *Mersennus*? A rayling Satiricall Babler, not able to make a reply in his owne defence, and therefore being put to a *Non plus*, hee went like a second *Job* in his greatest vexation to aske Counsell of the learnedst Doctors in *Paris*: And at last for all that, he fearing his causes, and finding himselfe insufficient, procured by much intreatie his friend *Peter Gassendus* to helpe him, and called another of his friends unto his assistance, namely, one *Doctor Lanovius*, a semenarie Priest, as immorall as himselfe, and one that professeth in his iudiciary letter much, but performeth little. And in good faith, I may boldly say, that for three roaring, bragging, and fresh-water Pseudophilosophers, I cannot parallel any in Europe, that are so like of a condition, as are *Mersennus*, *Lanovius*, and *Foster*.[53]

Gassendi, on the other hand, Fludd found to be a good philosopher, "an honest and well conditioned Gentleman, just as well unto his Adversary as friend, not passing beyond the bounds of Christian modestie, but striking home with his Philosophical arguments, when he seeth his occasion."[54]

To the charge that Foster repeated about Mersenne's wonder that King James "would suffer such a man to live and write in his Kingdome," Fludd tells of the close examination of his philosophy by James in person, after which, he received royal favors and patronage.[55] Foster also supposed that Fludd had published his books "beyond the seas" because Mersenne had accused him of being a magician; but Fludd replied with some interesting information about his publishers:

I sent them beyond the Seas, because our home-borne Printers
demanded of me five hundred pounds to Print the first Volume,
and to find the cuts in copper; but beyond the Seas it was
printed at no cost of mine, and that as I would wish: And I had
16 coppies sent me over with 40 pounds in Gold, as an
unexpected gratuitie for it.[56]

It was also galling to Fludd that the Anglican Foster had relied
on "Jesuites, Fryars and Seminarie Priests" as his sources to bring
aginst the weapon-salve and Fludd as a magician, since both
Englishmen were considered heretics among that company.[57] The
charge of heresy Fludd attributed to spite because of the popularity
of his books on the Continent. He then points out that even
Mersenne offered to join with him, if he converted: "But if you
Robert Fludd will leave your Heresie, I with my friend will heartily
embrace you, and will eyther face to face speake with you, or by
letters conferre with you about certaine Sciences, and I will desire
him not to write against you."[58] And elsewhere Mersenne offered
that if Fludd would leave his heresy, he could join Mersenne and
his friends in the "correcting of Arts," and told of the applause he
would receive from the entire commonwealth.[59] That brought up
another sore point with Fludd:

This I speake to some of my Countrymens shame, who instead of
encouraging me in my labours (as by letters from many out of
Polonia, Sueuia, Prussia, Germanie, Transylvania, France and *Italy* I
have been) doe prosecute me with malice & ill speeches with
some learned *Germans* hearing of, remember mee in their letters
of this our Saviour Christ his speech: *No man is a Prophet in his
owne Country.*[60]

As Fludd's statement above and Mersenne's correspondence
show, Fludd's debates with his adversaries were widely followed
throughout the learned European community.[61] It is also clear that
scientific, philosphical and religious convictions at this point were
still well bound together, even among those who professed their
separation, and that motives for opposition to Fludd were not
purely for scientific reasons. It was ingenuous of Mersenne, for
example, to call Fludd an evil magician on the one hand, but to
welcome him as a colleague on the other if he would but renounce
interest in, alchemy, natural magic, Platonism, astrology and re-

modern observers to paint the figures of the age simply in black or white; by praising the pioneers of the Scientific Revolution for their modern discoveries, while ignoring or dismissing their ties to, or interest in, alchemy, natural magic, Platonism, astrology and related areas. Kepler, Harvey and even Newton are obvious examples. Their scientific discoveries emerged from, and were clearly shaped by, their philosophical and religious framework. Fludd stood out because he *emphasized* the philosphical/religious scheme and not his practical achievements, such as medical skill and steel-making, which arose from experimentation and experience. As an intuitive Neoplatonist type, this is exactly what one would expect; but it certainly should not detract from his importance in the context of the times.

What Fludd did for the greater and lesser cosmos in the massive *Utriusque cosmi . . . historia*, he did for medicine in the *Medicina Catholica*, which was issued in four parts from 1629 to 1631 (see Bibliography). This equally massive work, which runs to some 1250 folio pages in all, deals with the causes and treatment of disease in overwhelming detail, and contains many excellent copperplate engravings to illustrate the text. The heavenly bodies, the winds and earthly elements all affect human well-being, and these influences are spelled out at great length. A separate tract deals with the pulse, and how it fits into his harmonic macrocosm-microcosm scheme. The dedications of this work also testify both to Fludd's standing with and to his acceptance in the highest ranks in England: the first tract of the *Medicina Catholica* was dedicated to William Paddy, a respected colleague in the College of Physicians and physician to James I; the *Integrum Morborum Mysterium* to George Abbot, Archbishop of Canterbury; the *Katholikon Medicorum Katoptron* to Sir Robert Cotton, the noted scholar and antiquary with a famous library. The *Catholic Medicine*, with its high-ranking dedications, appeared during the height of the Mersenne–Gassendi and Foster confrontations, which indicates the strength of Fludd's convictions and the level of support he could count on in England. In his reply to Foster, Fludd noted that his works were registered in two university libraries.

> And surely, if my conscience had perswaded mee, that there had been any thing in them, which had beene so haynous or displeasant, eyther to the Kings Majestie, or the Reverend

69

Bishops, I would not have presumed, to have made first our late King *James* of blessed memorie, and next three of the Reverend Bishops of the Land the Patrons of them; being that I, electing them my Patrons, must present them with the first fruits, and therefore must know, that if any thing had happened amisse in them, it could not bee hidden from them, whom in veritie I would bee afraid to displease, as being such as with my heart I reverence.[62]

Fludd's last work was published posthumously in Latin (1638) and English (1659), and its title bore the final name for his particular synthesis of Renaissance Christian Neoplatonist elements: *The Mosaicall Philosophy: Grounded upon the Essential Truth or Eternal Sapience.* In the foreword to the "Judicious Reader," Fludd requested the reader not to judge his book harshly if he uses excerpts from Scripture as proof of his philosophical discourse, "seeing that the holy Bible doth fully handle and set down the subject of both . . . sciences, . . . namely Physicall and Metaphysicall."[63] He then proceeds to demonstrate the truth of his propositions and the falseness of the Aristotelian by using the weather-glass, and then goes on in three hundred folio pages to deal with all the characteristic themes of his philosophy. At the end he excuses himself for the "roughness and harshnesse of my pen . . . and the insufficiency in the polished nature thereof," but the essential Philosophy does not need any "golden-tongued Oratour, nor smooth and methodical Rhetorician, or lip-learned Philosopher" to do her honor.[64] But on the other hand,

> the unworthy worldlings will not acknowledge or receive her with reverence, as they ought to do, but rather hide her perpetually, by their best endeavors, with the vail of obscure ignorance, and thereby do not desist to persecute and crucify daily that spirituall Christ, which is the onely verity, true wisdom, corner-stone, and essentiall subject of the true philosophy.
> Who onlely hath made heavens and earth, and every thing therein, and sustaineth and preserveth them by the vivification of his Spirit; which operateth all in all . . . who will defend his servants from the oppression of evill-minded men, and stand as a shield of defence, to preserve the proclaimer of his truth from the

70

Serpentine tongues of malitious backbiters, and the venemous carpings of the Cynicall and Satyrical *Momus*.[65]

These words were written at the end of a very productive life, but one that obviously had been far from free of strife. For over twenty years Fludd was acutely aware of the difficulty of having what he considered his most profound insights accepted broadly by the learned community; no doubt that's why he sympathized with the Brothers of the Rosy Cross. He would have had difficulty under the best of conditions, but at this particular time his intricate, interconnected edifice was about to become a monument to an era that the world was leaving. Many parts of Aristotle's science and philosophy were refuted in the Renaissance and its aftermath, but he is still admired for his attempts to organize and synthesize knowledge into an interconnected whole. Fludd is in a comparable position relative to Renaissance Christian Neoplatonism and all its centuries of antecedents. Although his science and other parts of his philosophy fell out of favor because of the major intellectual shift to a mechanistic view, they are no less a great achievement in creating a universe, totally complete and interrelated, in greater scope and elaboration than that of any age.

Chapter VI

FLUDD'S LATE RENAISSANCE CHRISTIAN NEOPLATONISM IN HISTORICAL PERSPECTIVE

> When philosophy paints its grey in grey, then has the shape of
> life grown old . . . it cannot be rejuvenated, but only understood.
> The owl of Minerva spreads its wings only with the falling of
> dusk.
>
> G. W. F. Hegel, Philosophy of Right

Taking a large view, the Renaissance as an historical moment may
be taken to include two major elements: the recovery of the art,
science, literature, religion and philosophy of the traditional
ancient civilized world, and a simultaneous movement toward the
modern one, driven by economics and supported by empirical
science. The second development became dominant when the first,
much of which was the product of ancient urban civilization, was
more or less complete. In fact, the modern world-view could not
make great strides until the preoccupation with the ancients ended,
and this could not happen until the task of recovering the revered
ancient wisdom had been nearly finished, and the findings had
been synthesized and absorbed by Western society. Only then was
it mature enough to begin to criticize the ancient authorities for
their inconsistencies and errors, and begin to gather and analyze
objectively its own data, and follow new leads, no matter how they
clashed with institutionalized belief.

The millennia of accumulated ancient knowledge, the golden
treasure of the truth-seekers of the early Renaissance, came from
the well-known centers of ancient Western Civilization: the Near
East (with much imported from the Far East): Babylonia,
Palestine, Egypt; Greece and its extension through the Hellenistic
period, followed by its medieval Islamic successors; and the Roman
civilization. Each originated, borrowed from others and transmitted
ideas from their time and place to succeeding generations until, in
the West, the high urban Roman civilization devolved into myriad
rural areas struggling for survival from famine, disease and
conquest. When, some five centuries later, Western Europe re-

72

emerged around commercial urban centers, the ancient world as it had been known was all but dead except in name. The fiction of a "Holy" Roman Empire, centered in Germany, reappeared, and those living at the time did not consider that there was any discontinuity between their time and the ancient Roman era. But in fact it was a new society, the beginning of the modern one. The feudal institutions, invented as a means of survival, took root to become, except in Germany and Italy, the basis for strong central monarchies which evolved into nation-states. Universities, unique to Western Europe in the High Middle Ages, came into being to serve the needs of the new urbanized society, and competitive commerce and international banking accompanied the rapid growth of population. Indeed, a totally different culture had been created with its own art, architecture, literature, music, education and religious practice.

In the middle of the fourteenth century, a time of economic depression, famine and plague following the boom of the High Middle Ages, the pioneer humanist Petrarch conceived the idea that there had been a break, a "middle age," between the glories of ancient Greece and Rome and his own time, which he characterized as a time of religious, political and cultural division and decay. In order to regain those past glories in a unified, revitalized society, it was necessary to return to direct study of the ancients, free of their medieval accretions, distortions and omissions, and apply their lessons by example to late medieval society. This increasingly powerful idea was transformed into a cultural ideal and put into action in the relatively free (from Papal and Imperial domination) commercial cities and some princely courts of Northern Italy, centering in Florence. The flowering of the Italian Renaissance in the fifteenth century, and its subsequent adoption in the North, took place with two of its seemingly contradictory currents continually at work: a looking back for wisdom and inspiration, accompanied by a concurrent building of modern institutions, which was aided by new methods, new discoveries and new technology. Thus the Renaissance was at once the revival of the ancient world and the forming of a new society which would lay the former to rest.

In looking back, Renaissance scholars also fully re-encountered the two great conflicting schools of thought, the Platonic and the Aristotelian, both of which would eventually be replaced by the modern mechanistic view during the seventeenth century. The paths taken by the legacies of the philosophic giants of classical Greece for the two thousand years between fifth-century B. C. Athens and fifteenth-century A. D. Florence were many and complex, and each acquired admixtures, extensions and alliances through their varied interpreters – philosophic, religious, scientific and medical – along the way. Since in the Renaissance both traditions attracted even more affiliations and interpreters, who often did not agree among themselves, modern scholars have frequently despaired over the inadequacy of the terms "Platonism" and "Aristotelianism" to cover the great variety of thought of the time. For example, a recent study refers to "Renaissance Aristotelianisms,"[1] and perhaps a full description of Robert Fludd's great synthesis could be labeled as Gnostic-Pythagorean-Platonic-Neoplatonic-Hermetic-Cabalistic-Judeo-Christian with some Aristotelian elements. I have normally shortened this list to call Fludd's work Renaissance Christian Neoplatonism. However, for the purpose of the broad view, the richness of these traditions (the Platonic and Aristotelian) may be reasonably represented by the commonly used terms because of the distinct character of their key elements and the kinship of those elements with allied thought.

Much of the problem with definition begins with Plato's writings themselves, which come down to us in some twenty-eight dialogues, of various lengths, and twelve letters. They are extremely broad-ranging (perhaps thus Platon, "broad") and show their author's mastery of the extant literature, folklore, sciences, medicine and mathematics of the time. Plato (428–348 B.C.), who was named Aristocles by his noble Athenian family, was well-educated and apparently traveled widely throughout Greece as well as to Egypt, Sicily and perhaps other areas. As a young man, Plato became a student of the legendary Socrates, who was executed in 399 upon conviction on the charges of corrupting the minds of Athenian youth and not accepting the traditional gods. Thus the bright young man who had seen, heard and read so much had his wits continually challenged and sharpened by Western civilization's premier teacher, with the result that after he founded his Academy, about the age of forty, and committed his philosophical works to writing, Socrates and his method of analogy, allegory and progressive questioning played a prominent role in many of the major dialogues, as was also true for the training at the Academy.

74

Another result was that Plato's dialogues do not contain a formalized, systematic philosophy. They do, however, cover basically the entire range of human thought and spell out positions on some philosophical issues clearly. Plato's original speculations drew on important predecessors and contemporaries, taking in particularly the number-mysticism and the universal mathematical harmony of nature from Pythagoras.

Yet many questions are raised and debated in depth, but apparently left unresolved, quite deliberately. Plato's genius was that he combined a keen intellect, capable of philosophical rigor of great complexity and subtlety, with the soul and expressive powers of a poet. Since his artistic side held that truth lay beyond the world of sense, the dialogues are a teaching device to help students seeking that truth to find it individually by going beyond themselves, and therefore the dialogues often delineate a process, not an end product: "The Platonic dialogues do not represent a doctrine; they prepare the way for philosophizing. They are intended to perform the function of a living teacher who makes his students think . . . "[2]

Such an open-ended set of writings, which covers a great range in masterfully executed style, obviously lends itself to be interpreted and extended in a variety of ways, not unlike the great books of the major religions. The *corpus platonicum* would be particularly attractive to those who share a like-minded approach to certainty, which could include some of the main tenets of Platonic sentiment, e.g., that the material world is not the ultimate, most real and most perfect one, but a corrupt and imperfect, ever-changing reflection of the perfect, unchanging world beyond that of sense; that there is a Creator who made the universe in a geometric and intelligible fashion; that the soul is immortal and returns to the other world following bodily death, where, unless corrupt, it encounters things in their purest form; and that the other world can be reached in this life through an inward striving, a dialectic process which takes one up to the point where a mystical connection and enlightenment may be achieved. Those who have sucessfully attained this enlightenment, the true philosophers, would not only be the teachers and advisors, but also, despite their natural reluctance, the rulers. By extension of these concepts, knowledge of the mysteries of the other side of things, the inner workings, as it were, also carries with it the possibility of mastery over physical phenomena and their manipulation, that is, natural magic, with all the power that is implied in such knowledge.

On the other side of the coin, we have Plato's famous student, Aristotle (384–322 B. C.), the sometime tutor to Alexander the Great, who departed radically from his teacher's ideas in rejecting an immortal soul, creation, and a perfect otherworld in favor of the sensible world as the only "real" one. Instead of an idealism which devalued the material world and strove toward the perfection only to be found in the non-material one, Aristotle set his sights on logically analyzing and systematizing the one we live in, with monumental results. His voluminous extant writings, which reflect the technical content of the lectures in his school at the Lyceum (his earlier, popular works, including dialogues, were lost in late antiquity), are encyclopaedic in character and include treatises on logic, physical sciences, biology, psychology, metaphysics, ethics, politics and literary arts. Here, in complete contrast to Plato, is a set of empirically-based, highly detailed, logically analyzed and systematic tracts which contain much with practical as well as theoretical application.

The philosophies of Plato and of the mature Aristotle may be envisaged as a relationship of thesis to antithesis, and their continual interplay, part in contrast, part in synthesis, constitutes a fascinating historical pageant up to the end of the Renaissance.[3] Although both schools continued after the deaths of their founders, the decline in importance of Athens as a city-state, with a corresponding rise in importance of the cosmopolitan regional polity of the succeeding empires of Alexander and Rome, caused the two great competitors to fade in favor of such schools as the Stoics, Epicureans, Sceptics and Eclectics. But about the beginning of the Christian era, there was a dynamic revival of earlier Greek philosophy, which lasted until the end of the Roman Empire in the West in the fifth century and had a major impact on science, medicine and religion, as well as philosophy, for more than ten centuries. Much of the revival was centered in Alexandria, Egypt, which became a crossroad between East and West for both commerce and ideas, and thus developed into the greatest intellectual center of late antiquity, with the finest library of the ancient world.

One of the important early movements in this revival, the Neopythagoreans, flourished in the first century A.D., among whom may be counted the legendary Apollonius of Tyana (died c. A.D. 97). Philostratus' *Life of Apollonius* (c. A.D. 217), written for the erudite wife of Emperor Septimus Severus, Julia Domna, established him as the prototype of a much-traveled seeker of truth who gains great ancient wisdom in Babylonia, India and Egypt, and is thereby able to work many wonders, including curing the sick and divination. He was, as a result, accused by some of being a magician and sorcerer, which is another common characteristic of figures in this tradition, just as Robert Fludd and the Rosicrucians would be accused a millennium and a half later. Typically for the age, the Neopythagoreans were eclectic in their borrowings from Pythagoras, Plato, Aristotle, the Stoa and perhaps the Gnostics, as well as from some Near Eastern traditions. It was this syncretic pattern which provided influences for a number of important movements of the time, and became the reason why Renaissance Platonists found the possibility of combining various religious sentiments and the philosophy, cosmology, medicine and science of earlier writers into a comprehensive whole.

Although there was at the time no separate Aristotelian school, two of the best and longest-lived examples of syncretic revival which fall into this tradition are found in the works of Claudius Ptolemy (c. A.D. 90–c. 168) and Galen (A.D. 131–201). Ptolemy was educated in Alexandria, where he gathered together the diverse observations of the heavens by Aristotle, Hipparchus (190–120 B.C.), Posidonius of Rhodes (c. 135–50 B.C., an eclectic Stoic who combined Plato and Aristotle) and others, and added his own into the well-known Ptolemaic system of cosmology in thirteen books called by its Arabic name, *Almagest*. In addition, he wrote an updated geography and the *Tetrabiblos*, or four books on the influence of the stars, an astrological primer. For fifteen centuries Ptolemy's incorrect, earth-centered, but reasonably accurate system was the standard, and it also earned the official approval of the church. The geometrically-based system of perfectly circular orbits and uniform speed of the planets, though derived from Aristotle and his predecessors, fits neatly into a Pythagorean/Platonic scheme as well.

Galen, the most prominent physician of antiquity after Hippocrates (460–370 B.C.), was a Greek born in Pergamon in Asia Minor who studied in his native city, Smyrna, as well as in Corinth and Alexandria in Ptolemy's time. He was well versed in all the extant philosophical schools and medical knowledge, and practiced medicine for a time in Rome. His voluminous medical legacy, the absolute standard until the sixteenth and seventeenth centuries, combined much of ancient medicine, Aristotle's natural philosophy, his own observations, and Pythagorean number-lore into a comprehensive medical encyclopedia. His works include two on astrological medicine (which Hippocrates insisted was absolutely necessary for a proper diagnosis), one of which was titled the *Prognostication of Disease by Astrology*. Though he was attacked in the Renaissance by the anatomists and Paracelsians for his faulty observations and Aristotelian bent, there was much in Galen that would filter through to Robert Fludd's medical theory and practice.

Another example of cross-influence to emerge in the melting-pot of Alexandria may be seen in the work of the Jewish scholars, the most important of whom was Philo of Alexandria (c. 25 B.C.– c. A.D. 40), whose mystical interpretations of the Torah foreshadow the later development of the Cabala. In medieval Spain, where Christian and Jewish ideas freely intermingled as they had in Alexandria centuries before, we find:

> Two different branches stemming from the same root meet in the doctrine of the Torah as it finally took shape in the Zohar. The ancient root is undoubtedly Philo of Alexandria, to whom we may ultimately attribute all these distinctions between literal meaning and spiritual meaning which were taken over by the Church Fathers and the Christian Middle Ages, and also by Islam.[4]

Thus the interconnections of these branches meet again in the Middle Ages and during the Platonist revival in Renaissance Florence.

In addition to the Neopythagoreans of the early Christian era, there were eclectic writers, now called Middle Platonists, such as Plutarch (A.D. 46–120), who wrote the famous *Lives of Notable Greeks and Romans*, and Apuleius (born c. A.D. 125), but no comprehensive and unified system emerged from this diverse group. There was another very influential collection of religious/ philosophical writings to come from Alexandria, however, about the secondary century A.D., which were attributed to Hermes Trismegistus, "Thrice-Great" Hermes. The collection of books written under that name, the *corpus hermeticum*, was thought to be very ancient, and is discussed further below with Renaissance Platonism. The Hermetic writings follow the syncretic Platonic-based pattern of the time, and include Gnostic, Stoic, Neoplatonic and Christian elements. They are most notable for their influence on the early Church Fathers and the Renaissance Platonists.

A unified system arose from the Neoplatonists, the successors of the Neopythagoreans and the Middle Platonists. The pioneer Neoplatonist was the Alexandria-educated Plotinus (c. 204–c. 270), whose collected works, the *Enneads*, and biography were compiled by his student Porphyry of Tyre (c. 232–c. 302). In Porphyry's own books and teachings, he continued the process started by his master and predecessors of incorporating diverse but related doctrines (including Orphic and Chaldean works) into an increasingly interconnected system of emanative stages between God and man which may be traversed from within, culminating in a mystical union. One of Porphyry's important students, Iamblichus (died c. 330), who taught in Syria, greatly proliferated the number of beings in the hierarchy, but the whole Neoplatonic scheme was given its most complete and systematic expression by Proclus (410–485), who taught in Athens. His *Elements of Theology, Platonic Theology* and commentary on key Platonic dialogues provide, unlike the *Enneads* (which were a collection of essays and lectures), the one true structurally unified exposition of Neoplatonic thought.

In Alexandria, the Neoplatonism which began there in the third century under Ammonius Saccus, Plotinus' teacher, was absorbed in many places by the Church Fathers, who borrowed concepts and terminology for fleshing out the still-developing Christian theology. Thus Clement, Eusebius and St Augustine (who had been, among other things, a Neoplatonist pagan before his conversion) all used the Neoplatonist tools at hand to help fashion the philosophical framework for the developing religion. Another Platonist-related Christian movement that flourished in the second century A.D. in North Africa and elsewhere was that of the Gnostic Christians. The origins of their doctrines in pre-Christian times are not known; but their dualism, system of emanations, *gnosis* (which is at once self-knowledge and knowledge of God, which provides the means of salvation) and reincarnation would seem to borrow from a variety of traditions, East and West, including Neopythagoreanism and Neoplatonism. Little was known about the heretical gospels and other writings that circulated among them until recently, except by way of the polemics against them by orthodox Christians, particularly the one written by the Church Father Irenaeus, Bishop of Lyon (born c. A.D. 140), about the year 180.

In the last few years, the 1600-year-old Nag Hammadi Gnostic manuscripts, which were written in a Coptic translation of the original Greek and discovered in 1945 in Upper Egypt, have been published in the original and in English translation. Included in this fascinating set of varied Gnostic and other tracts are excerpts from Plato's *Republic*, and two Hermetic works, the *Discourse of the Eight and Ninth* and *Asclepius*.[5]

Thus was the Mediterranean world of the Roman Empire rich in Platonist-related movements which had freely borrowed from the many traditions available to them in this time of wide intellectual exchange. Aristotle, though not studied as a separate school during these centuries, was well known because of the insistence of the Neoplatonist and similar movements on integrating his writings into a synthesis with Plato. Of particular importance was the appropriation of his logic and natural philosophy, which would be passed on in part to the Middle Ages through this medium.

Three figures of late antiquity were the most important for transmitting the thought of Plato, the Neoplatonists and some of Aristotle to the succeeding Middle Ages in Latin Europe: St Augustine, Psuedo-Dionysius and Boethius. By virtue of his widely influential work, Augustine (354–430), the greatest of the Latin Church Fathers, is particularly important in this context because of his incorporation of Platonic and Neoplatonic thought into Christian doctrine. Born to a Christian mother and a pagan father in what is now Algeria, Augustine was educated in Tagaste, his birthplace, Madaura and Carthage. After rejecting Christianity as a young man, he became a Manichaean and, in 384, went to Rome and then to Milan in order to teach rhetoric. During this period, he read some Neoplatonic works, probably Plotinus' *Enneads*, which gave him the intellectual framework he needed to eventually accept Christianity. He subsequently converted in 386, and was baptized the following year by St Ambrose. In 391, Augustine was ordained by the Bishop of Hippo in North Africa, and assumed that office himself from 396 until his death in 430. His literary output was prodigious, and although he did not have a successor school and students, and his town was overrun by barbarian invaders at the time of his death, his works were widely read and broadly influential in the ensuing centuries.

Because of his personal experience, Augustine insisted that "none of the other philosophers has come so close to us as the Platonists have,"[6] and therefore one could be led to Christianity through them, as he expressed it in the *Confessions*: ". . . after reading these books of the Platonists which taught me to seek for a truth which was incorporeal, I came to see your invisible things, understood by those things which are made."[7]

Later, in reading St Paul, Augustine "discovered that everything in the Platonists which I have found true was expressed [there]."[8] As with some of the other Church Fathers, he speculates that Plato may well have encountered the Scriptures in his travels and study in Egypt, which could account for the many points of similarity. In the Renaissance, Augustine's suggestions about Plato's connections with the Mosaic books, as well as those of other Church Fathers, were picked up and amplified by the Florentine Platonists and later followers of this tradition, including Robert Fludd.

Whereas Neoplatonism led Augustine to Christianity and was closest to that religion of all extant Western philosophies, in his view, the several works circulated under the name of Dionysius the Areopagite (St Paul's Athenian convert) from about the end of the fifth century fearlessly synthesized Christianity and Neoplatonism. The real author, possibly an unknown monk who lived in the Greek Eastern Empire, brought great attention to his tracts by attributing them to someone of importance, a common practice of the time. His longer works, *On Divine Names*, *On Celestial Hierarchy* and *On Ecclesiastical Hierarchy*, as well as the smaller but influential *Mystical Theology*, are here well summarized:

> The principal parts of the Dionysian legacy to the West were the negative and superlative theology, the strongly hierarchic conception of being, an elaborate angelology, and the doctrine of mediated illumination and spiritual knowledge conferred by the sacraments and by angelic ministration.[9]

Derived from the Church Fathers – Gregory of Nyassa, Clement of Alexandria and Origen, and the Neoplatonists Proclus and Plotinus – Pseudo-Dionysius was appealed to as authentic as early as the Church Council of 533, and was translated into Latin in 858 by John Scotus Eruigena (who also wrote a commentary) for the Frankish king, Charles the Bald. The writings were thereby made available to medieval scholars, and were commented on by such figures as Hugh of St Victor, Robert Grosseteste, Albertus Magnus and Thomas Aquinas, all of whom considered them authentic. Thomas Aquinas and Dante added their own ideas about the compositon of the celestial hierarchies, and both Marsilio Ficino, who translated the Dionysian works anew in the Renaissance, and his student Pico della Mirandola made extensive use of this doctrine. The importance of this system of Christian Neoplatonist celestial hierarchies and other Dionysian tenets for Robert Fludd is everywhere manifest in his works.

Along with Augustine in North Africa and Pseudo-Dionysius in the East, Boethius (480–525), the last notable intellectual figure of ancient Rome, completes the circle of the late antique Mediterranean world as a major legacy to the medieval one. He was born of a noble Roman family at the time of the last of the Roman emperors in the West, and rose through the Senate to the office of consul by 510, after which Theodoric the Ostrogoth entrusted him with the administration of domestic affairs and the foreign policy of his kingdom. For some still unclar reason, after a year in office he was imprisoned and subsequently executed for treason. What distinguishes Boethius is not so much his political career, but the astonishing scope of his scholarly undertakings: the translations of Aristotle's *Logic* and Porphyry's *Isagoge*; various commentaries on Aristotle, Porphyry and Cicero; and his own original tracts, particularly the *Consolation of Philosophy*, which was written in prison.

Boethius was eclectic in his thinking in that in metaphysics and cosmology he agreed with Plato, and in natural science and logic with Aristotle, and believed there was no conflict between the two. In fact, prefiguring Giovanni Pico della Mirandola in Renaissance Florence a thousand years later, he laid out for himself the impossible task of translating the entire works of both Plato and Aristotle into Latin, and then demonstrating the reconciliation of the two systems. Although he did not nearly achieve this monumental task, what he did accomplish was both impressive and influential. Among other important contributions to the Middle Ages was the translation of the *Logic* of Aristotle, which became basically all that was known of him until about the twelfth century, and the equally important application of Aristotelian logic to theology. This was combined with the coinage of Latin philosophical terms and the composition of commentaries, a form of writing which became very popular in subsequent centuries. Rightly, Boethius has been called the last of the ancients and the first of the medieval schoolmen, for he indeed lived at the clear transitional point between these two ages.

And so it was that Augustine and Pseudo-Dionysius were primarily responsible for a Platonic Christianity being transmitted to the following centuries, while Boethius, foreshadowing Thomas Aquinas, brought a strong dose of Aristotelian logic to Christian theology, which was mixed together with Platonic metaphysics. Thus as purely pagan scholars passed from the scene and Christianity became dominant everywhere in the West, the two major philosophical schools became thoroughly enmeshed with Christian doctrine. Other authors of late antiquity who should be noted as contributing to this tradition include the fourth-century writers Chalcidius, whose translation of, and commentary on Plato's *Timaeus* provided the Middle Ages with almost all of its direct knowledge of his dialogues, and Marius Victorinus, who translated Porphyry's *Isagoge* and some of Plotinus' *Enneads*. From the fifth century, Martianus Capella's liberal arts compendium, called *On the Nuptials of Mercury and Philosophy*, should be noted, and in the sixth century Boethius' student Cassiodorus instituted the division of the liberal arts into the *trivium* (grammar, rhetoric, dialectic) and the *quadrivium* (arithmetic, geometry, music, astronomy), a usage which remained for the next millennium and a half. (This was the basis for Robert Fludd's Oxford education in the sixteenth century, and figured in a complaint of his in the seventeenth over the content of these subjects.)

During the Middle Ages, Plato and Aristotle, in the forms passed on by the Hellenistic and Roman worlds, traversed three main paths: the Latin West in the territory of the collapsed Roman Empire; the Greek Eastern or Byzantine Empire; and the Arabic-Islamic world, which expanded to include the Near East, North Africa and Spain. Great conflict between the two philosophies arose only in the Latin West, primarily because Aristotle had been little known there for some five centuries, and was reintroduced with great impact. In the Byzantine Empire, the complete works of Plato, Aristotle and the Neoplatonists were maintained and studied, but it was the Neoplatonist standpoint that prevailed by comparison with the other two. The Arabs, on the other hand, were much more interested in science (including astronomy/astrology and alchemy), mathematics and medicine, and translated nearly all of Aristotle's work into Arabic along with Neoplatonist commentators (which influenced their understanding of him); the latter included the apocryphal *Liber de causis* and *Theologia Aristotelis*, which were actually derived from Proclus and Plotinus respectively. Plato himself was relatively neglected, however, with only a few dialogues deemed to be sufficiently interesting or important to translate.

In the Latin-speaking West, during the first five centuries of the Middle Ages, study of the Greek philosophers, and indeed, philosophy in general, suffered greatly, and knowledge of Greek all but disappeared, except in some isolated monasteries, primarily in Ireland. With the exception of John Scotus Eruigena (c. 810–c. 877), there were no original philosophers of note in Latin Europe during this time, but the short-lived Carolingian Renaissance greatly aided later scholars by their masterful copying of ancient documents, many of which subsequently perished in the original.

Because of its nature, in both method and content, and the timing of its reintroduction on a large scale in the West, beginning in the twelfth century, Aristotelianism became institutionalized in the newly-invented universities and eventually also in the Roman Church, while Platonism, when reintroduced in force in the Renaissance, would find itself in a not unfamiliar anti-establishment role. When, about the beginning of the twelfth century, Western Europe began a two-century transformation from a culturally backward, agrarian society ruled by a fractious warrior class, it became a more dynamic, stable, urbanized, commercial society which eventually developed a higher cultural level than its Byzantine and Arab neighbors. Instead of being constantly under attack from the Norsemen, Saracens and Magyars, Westerners went on the offensive with the Crusades, the Christian reconquest of Spain and the Norman conquest of Sicily and Southern Italy. While the Crusades stimulated commerce and trade between East and West, the Sicilian, Southern Italian and particularly the Spanish conquests opened the Arabic world of learning to Western scholars, who began to translate many unknown or forgotten works of antiquity into Latin. Most of the new translations were works by

Aristotle, and they included his Islamic interpreters, most notably Avicenna and Averroës. What was particularly significant about the translations coming at this time and place was that the universities, which were church-run institutions, developed to meet the needs of a new society that the old monastery schools could not meet, had already incorporated such parts of the logic of Aristotle as were available to them. Using this logic as a foundation, they had developed a dialectic or scholastic method of study for all fields (except revealed religion) that was given its most used form by Peter Abelard in his *Sic et Non* (c. 1122). When the mass of nearly the entire extant *corpus* of Aristotle, with learned commentary, burst upon the universities, it was for the most part eagerly received, but it created both philosophical and religious problems. Not only did some parts of Aristotle clearly clash with basic Christian doctrine (particularly on the immortal soul and creation), they also clashed with other Platonist elements incorporated centuries before, and carried forward by the followers of Augustine and most Franciscans. Thus ensued such controversies as the primacy of reason or faith. However, at the most important seat of higher learning in Europe of the High Middle Ages, the University of Paris, the works of Aristotle, some of which were condemned early in the thirteenth century, eventually became the standard encyclopedia of study, and by the mid-fourteenth century all Arts graduates were required to demonstrate a knowledge of the known works in use. Aristotelian-based studies spread to the other universities as well, since both the logical method and the wide-ranging content were well suited for institutional instruction. For the church, the adoption of Aristotle was aided by the oustanding reconciler of Aristotelianism and Christian doctrine, St Thomas Aquinas. His *Summa Theologica* (1265–72), other works and commentaries on Aristotle not only cleared the way for Aristotelian-based thought to become eventually the official doctrine of the church, but also served to strip away much of the accretions from Neoplatonist, Arabic and other sources, so that the differences between the two great philosophic systems could stand in greater contrast. Although the Thomist synthesis was not at first accepted by the conservative theologians, and was partly condemned in 1277, it gained a following among the Dominicans, and by the time of the Jesuits and the Council of Trent in the sixteenth century, it became the principal weapon against the arguments of the reformers, starting with Luther (who was, interestingly, an Augustinian monk).

On the other side, the greatest contemporary foe of Thomism was the Franciscan Duns Scotus (c. 1265–1308), who was in turn attacked by William of Ockham (c. 1285–1349) with his famous nominalism, but the fourteenth century also saw the development of two other movements with major impact in the ideological wars. The first was that of the Christian mystics, the main figures of which were the German Dominicans Meister Eckhart (c. 1260–1327), Blessed Henry Suso (c. 1295–1366) and John Tauler (c. 1300–1361), and the Flemish mystic Jan van Ruysbroeck (1293–1381). The latter's speculations on the mystical experience had clear Neoplatonic affiliation.

The second, and potentially greater, movement began with Francesco Petrarch, the "father of humanism" and conceiver of a "dark" or "middle" age lying between his own time and that of ancient Rome. Petrarch not only started a movement to look back to the ancients in general as sources of wisdom for his society, but to Plato in particular. Although he could not read Greek, much to his regret, one of his prized possessions was a manuscript of Plato in its original language (he also owned all the then extant Latin translations), and he readily invokes Plato's name against the followers of Aristotle. Petrarch greatly admired Cicero (who had also been enamoured of Plato), as well as St Augustine, and considered Plato to be the "greatest of Philosophers." Thus was a lively interest in the works of Plato in particular fostered in the embryonic Italian Renaissance.

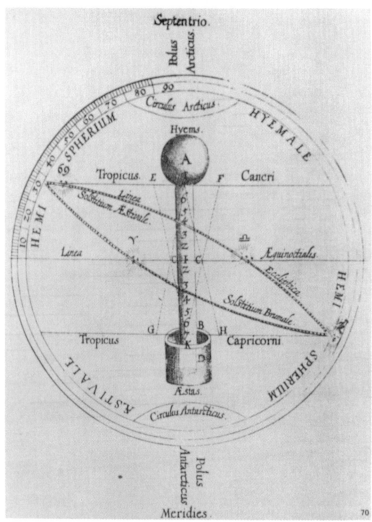

10. The Weather-Glass and its Correspondence with the Earth
(MC I, 2, i, p. 28)

11. The Winds and Their Governors (MC I, 1, p. 113)

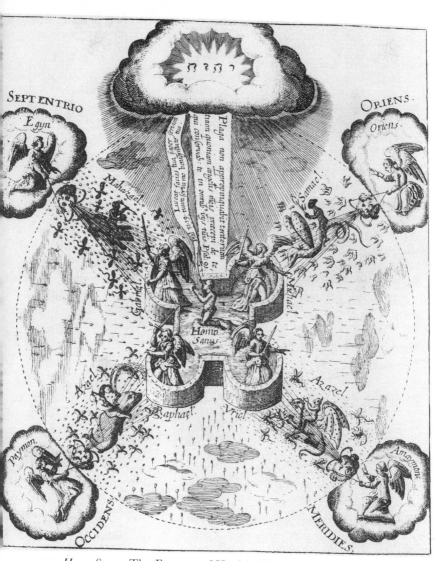

12. *Homo Sanus*. The Fortress of Health. The Angels Repel the
Onslaught of Evil Forces (MC I, 2, i, p. 338)

13. The Invasion of the Fortress of Health. The Physician
(Fludd) Looks in the Direction of the Invasion for a Cure (MC
I, 2, i.): (3–4)

14. The Physician (Fludd) Takes Pulse of a Patient. *Integrum Morborum Mysterium* (MC I, 2, i, title page)

15. The Physician (Fludd) Casts Horoscope of Patient
(MC I, 2, ii, p. 253)

16. The Physician (Fludd) Examines Urine of Patient
(MC I, 2, ii, p. 253)

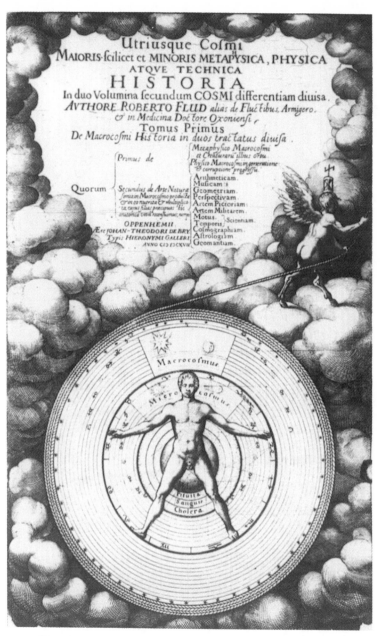

17. Title Page, *Utriusque cosmi ... historia*. The Microcosm and
Macrocosm, put in Motion by Time

18. Title page, *De Naturae Simia*. Depicts Man's Arts, through which He becomes the Ape of Nature

TOMI SECVNDI
TRACTATUS PRIMI,
SECTIO SECUNDA,
De technica Microcofmi hiftoria,
in
Portiones VII. divifa.
AUTHORE
ROBERTO FLUD aliàs de FLUCTIBUS
Armigero & in Medicina Doctore Oxonienfi.

19. Title Page, *The Microcosmic Arts* (UCH II, 1, ii)

20. Portrait of Fludd, *Integrum Morborum Mysterium* (MC I, 2, i, frontispiece)

21. Fludd's Monument in Bearsted Parish Church

In Iesu qui mihi omnia in tuta morte resurgam
VNDER THIS STONE RESTETH THE BODY OF Ro
BERT FLVDD DOCTER OF PHISICKE WHO CHAN
GED THIS TRANSITORY LIFE FOR AN IMORTALL
THE VIII DAY OF SEPTEMBER A.D. MDCXXVII
BEING LXIII YEARES OF AGE, WHOSE MONVMENT
IS ERECTED IN THIS CHANCELL ACCORDING
TO THE FORME BY HIM PRESCRIBED

22. Grave Marker in Floor of Bearsted Parish Church

23. Fludd's Signature on a Manuscript in the Bodleian Library

Chapter VII

FLUDD AND THE PLATONIC RENAISSANCE

It appeareth . . . that Plato had knowledge of the Word, and had read the Books of *Moses*; and for that reason was called, *Divinus Plato, the divine Plato*. In like manner, the excellent philosopher *Hermes*, otherwise termed *Mercurius Trismegistus* . . . was not only acquainted with *Moses* his books, but also was made partaker of his mysticall and secret practise . . .

Fludd, *Mosaicall Philosophy*

In the fifteenth century, which witnessed the full flowering of the Italian Renaissance, the interest in recovering classical authors was translated into an increasingly successful achievement with profound impact. Aristotelian studies continued to remain strong during the Renaissance as a continuation of the medieval incorporation of his works as the basis for instruction in the universities and monastery schools and in fact were strengthened by the new humanist translations and authoritative editions in the original Greek of the fifteenth and sixteenth centuries. But it was in the fifteenth century that the full works of Plato and the Neoplatonists, including Hermes Trismegistus, became known for the first time in the West since antiquity in either Latin or Greek, and could thus form the basis of a major philosophical alternative to Aristotle in philosophy, theology and science.

Early in the century, a small but growing group of humanists, who were trained in Greek by a Byzantine emissary in Florence from 1393–7, translated several of Plato's dialogues into Latin for the first time, and some of these became widely read. The following years saw a flood of manuscripts eagerly sought in the East and returned for translation.

In mid-century, Platonism was given a major boost by three figures involved in the Council of Ferrara-Florence (1438–1439), where the split between the Roman and Greek Churches was repaired after a four-hundred-year schism. Instrumental in arranging the council was one of the outstanding original scholars of the age, Nicholas of Cusa (1401–64). His Neoplatonist-based (but eclectic) philosophy and theology (influenced by both the German mystics and the Italian humanists) combined with his work in astronomy, dynamics, medicine, calendar reform and politics, mark him as moving out of his medieval scholastic training into the early Renaissance. Not only were his speculations influential in scientific, philosophical and theological circles, the theme of his life's work was the lofty goal of the Renaissance itself: unity and reform in the church, society and the underlying life of the spirit. Unfortunately, this goal would prove to be increasingly elusive in the ensuing full development of the Renaissance and the concurrent Reformation.

Two of the Byzantine participants in the Council, the aging scholar Gemistus Pletho and his student, Bessarion, who became a Roman Cardinal, kindled an increased interest in Plato and the Neoplatonists. Pletho's 1438 lectures in Florence left a lasting impression on the wealthy Florentine patriarch Cosimo de' Medici, and the subsequent 1453 Turkish victory over Constantinople, the last bastion of the Byzantine Empire, caused a great influx of Greek scholars and manuscripts into the West. In 1462, Cosimo decided to take a direct hand in making Plato, Hermes and the Neoplatonists available in Latin in a complete and systematic way, and to found a new Platonic Academy. He thus installed the bright young scholar Marsilio Ficino (1433–99), the son of a Medici physician, in a villa in Careggi, near Florence, with lifetime support and the mandate to devote his life to the coveted translations, and the study and teaching of Platonic philosophy. At the villa, Medici deposited the large store of Greek manuscripts that had been carefully collected in the East.

Over a period of thirty years, which was during the time of peace between the Italian city-states that permitted the Florentine Renaissance to flourish, Ficino carried out his patron's wishes. He translated the *corpus hermeticum* of Hermes Trismegistus (1462–63), the dialogues of Plato (1463–68), and the works of the major Neoplatonists: Plotinus, Porphyry, Proclus, Iamblichus and Pseudo-Dionysius (1484–92); he wrote commentaries on Plato, Plotinus and St Paul; and he penned three original works, the *Theologia Platonica* (1469–74), *De Christiana religione* (1474), and – on medicine – *De vita* (1489). The villa outside Florence was turned into a new Platonic Academy, where students and scholars could come to study and attend lectures, and thus it became the most important and influential center of Platonism since Alexandria fifteen hundred years earlier. Its Renaissance Christian Neoplatonist legacy eventually spread throughout the learned Western world, and clearly formed the basis for Robert Fludd's studies a hundred years later.

Even though all the necessary manuscripts had been assembled for Ficino to make the first Latin translation of the revered Plato's complete works, in 1463 Cosimo, who was nearing the end of his life, ordered that work on Plato should wait until an even more important set of manuscripts were translated, the fifteen-dialogue *corpus hermeticum* of Hermes Trismegistus. They were completed and read just before Cosimo's death the following year.

Dame Frances Yates, in her seminal work, *Giordano Bruno and the Hermetic Tradition*, was the first modern scholar to identify the importance for Renaissance thought of "Thrice-Great Hermes," who, for Cosimo de' Medici, took precedence even over the "divine Plato."[1] He was thought to be a very ancient Egyptian sage, probably a contemporary of Moses. For Renaissance scholars, the more ancient a source of wisdom the closer to its divine origin it was, and therefore the purer. Although the *Hermetica* were probably written by a Neoplatonist author or authors writing under Hermes' name about the second century A.D. in Alexandria, as previously noted, several of the early Church Fathers, including Lactantius, Clement, Origen, Eusebius and, particularly, St Augustine, regarded the dialogues of Hermes as well as the supposed works of Orpheus (the *Orphica*), Pythagoras (*Carmina Aurea*), the *Chaldean Oracles* (by Zoroaster, according to Pletho) and the Sibylline Prophecies as genuinely ancient. Since the Church Fathers were living in a time when Christianity was struggling for firm ground to be held against a number of strong pagan rivals, these ancient theologians (as they were regarded) and their tradition of *prisca theologia*, or ancient theology, were considered either to be precursors of Christianity or as having borrowed much of their theology and philosophy from the Mosaic teachings. This would, of course, account for the similarity in content of the authors, for example between the book of Genesis, the Hermetic dialogue *Pimander* and Plato's *Timaeus*.[2]

Lactantius, a Renaissance favorite, thought that Hermes was a major prophet anticipating Christianity, and was a very learned man of great antiquity who wrote many books filled with the truth of divine wisdom, and thus deserved the title Trismegistus. A very colorful description of the Hermetic works, indeed one which would easily capture the imagination of anyone interested in Hermetic lore, was given by Clement of Alexandria, even though it appears that he did not know the writings which came down to the Renaissance under that name. He portrays an ancient procession of Egyptian priests carrying forty-two books authored by Hermes: six pertain to medicine, and the remaining thirty-six encompass all of Egyptian philosophy. It is led by a singer who carries two books on hymns and music, and includes the *horoscopus* who bears four on the heavens. A place of mystery, initiation and magic is depicted, where the divine ancient wisdom, committed to a set of revered texts, is being solemnly carried in sacred procession. What wondrous secrets they must contain!

In addition to these two Church Fathers and mention of Hermes by others, there can be no doubt that the greatest weight given to the authenticity of Hermes came from St Augustine. Though Augustine specifically denounces the idolatry of the *Asclepius* dialogue in his *City of God*, which caused some minor concern among Renaissance Platonists, later in the same book he establishes Hermes as living in the third generation after Moses' time.[3] Considering the heavyweight imprimatur of Augustine that Hermes was a genuine ancient Egyptian sage, and the other lore attached to his name, it was little wonder that Cosimo pulled Ficino off the Plato manuscripts and had him work on the ones thought to be more ancient.[4]

But what then was the connection between Plato and Hermes? And how does a Renaissance Christian Platonist reconcile both with Christianity? Pletho may have started, or, at the very least, contributed to the Renaissance Platonists' various schemes dealing with these issues. Since the affinity between Hermes and Plato was clear, and Hermes was, according to unimpeachable authority, more ancient, it only remained to deduce how this sacred wisdom was passed down. Pletho believed that Zoroaster, whom he thought had written the *Oracula Chaldaïca*, was the first to receive the hidden knowledge, from whence it passed through a number of hands to Pythagoras and Plato, but he does not include Hermes. (Sacred wisdom was only to be passed down through select figures who had proven themselves worthy to receive it.) Ficino rectified this major omission by putting him at the head of the first of several proposed lineages:

He is called the first author of theology: he was succeeded by

Orpheus, who came second amongst ancient theologians:
Aglaophemus, who had been initiated into the sacred teaching of
Orpheus, was succeeded in theology by Pythagoras, whose
disciple was Philolaus, the teacher of our Divine Plato. Hence
there is one ancient theology . . . taking its origin in Mercurius
and culminating in the Divine Plato.[5]

Later, Ficino puts Zoroaster at the beginning of this line, then
modifies it to have Zoroaster and Hermes contemporaries in Persia
and Egypt respectively, and then carries it forward to Plato. Of
course, a continuation would lead through the Neoplatonists to
Ficino himself. In true Renaissance humanist style, he regards the
ancient tradition as having been broken after Proclus' school, and
then traces some surviving threads through the intervening
centuries: of the Byzantines, Psellus and Nicolaus of Methone;
from the Arabs and Hebrews, Avicebron, Alfarabi and Avicenna;
and of the medieval scholastic philosophers, Henry of Ghent and
Duns Scotus. In his own era, he notes particularly Bessarion and
Nicholas of Cusa for their Platonism, but includes Dante and
Guido Cavalcanti in the ranks. At the end of this long dark spell,
punctuated with occasional light here and there, stands Marsilio
Ficino himself, who had caused the light to shine brightly once
more.[6]

Indeed, Ficino had no doubt about his own historical position in
the Platonic tradition. As with many Renaissance humanists from
Petrarch's time onward, he had a keen awareness that his own time
was distinctly different from the preceding "middle" or "dark" age:

Our century, like a golden age, restored to light the liberal arts
that were nearly extinct: grammar, poetry, rhetoric, painting,
sculpture, architecture, music, the ancient performance of songs
with the Orphic lyre, and all that in Florence. And
accomplishing what had been revered among the ancients, but
almost forgotten since, it united wisdom with eloquence and
prudence with military arts . . . And in Florence, it restored the
Platonic doctrine from darkness to light.[7]

This short quote not only states succinctly the Renaissance
humanists' goals and accomplishments, but also shows the
importance Ficino places on the revival of Platonism in this
program. His own translations were seen as a rebirth or
resurrection of the neglected Plato, and his villa was a reopening of

the long-closed Academy; but it was not merely Plato and hi.
school that were revived. The significance was in the restoration, to
their minds, of a very ancient tradition of sacred wisdom which
was synthesized into and represented by Platonic doctrine, as
Ficino put it. This, of course, included not only Plato's teachings in
the dialogues and letters, but the Neoplatonic interpretations and
the *prisci theologi* (Hermes, Orpheus, Zoroaster, etc.) Thus the
Renaissance Platonism of Ficino was a broad synthesis of Plato and
the Neoplatonist, Pythagorean, Gnostic, Stoic and other elements
that were afield in the first centuries of the Christian era in the
Hellenistic-Roman world.

And what of the relationship between this pagan, divine wisdom
and Christianity? Can one be a follower of a long line of non-Judaic
theological tradition, which in the early centuries of this era was
embraced in part by some heretical Christian sects or even anti-
Christian ones, and still be an orthodox Christian in the
Renaissance? And what about the mysteries, initiation only of the
worthy, and the natural magic associated with the tradition? These
and other potentially troublesome questions were deftly handled by
continuing the Neoplatonists' process of syncretizing various
elements, thereby making a Platonized Christianity part of the
tradition.

Bessarion, for example, was convinced that Plato's travels in
Egypt put him in touch with the teachings of Moses, which was a
typical Renaissance notion of the interchange of sacred ideas,
including the lineage of initiations into the ancient sacred wisdom
from the *prisci theologi*.[8] For Ficino, the Platonist philosophy, which
he also considered a theology, was in complete accord with
Christianity and in fact confirmed its truths. In two short works he
attempted to demonstrate the harmony of the Mosaic and Platonic
doctrines, and drew a parallel between the life of Christ and that of
Socrates. Not only is Platonism a religious philosophy, but the
ancient philosophers in this tradition, which includes Pythagoras,
Socrates and Plato, were considered religious figures to the extent
that they achieved, as did the Old Testament prophets, eternal
salvation in their capacity as precursors of Christianity. As
examples of prominent Christian theologians, Ficino particularly
cites Bessarion and Nicholas of Cusa, but it is St Augustine whom
he quotes most frequently in defense of the compatibility of
Christianity and Platonism. The ancient Platonist theology plays

an even greater role than compatibility, however: according to Ficino, it actually enhances Christianity by bringing doubters or unbelievers who study it back into the faith (as it did for Augustine himself).[9] Platonism, therefore, plays a vitally important role in the divine world-historical plan, and Ficino follows this through by regarding himself as a key figure, an instrument, as it were, of the divine will in reviving the ancient theology as an oasis in a desert of Aristotelian non-believers:

> It was . . . by the will of divine Providence . . . that a religious
> philosophy arose among the Persians under Zoroaster and
> likewise among the Egyptians under Trismegistus, that it was
> then nursed by the Thracians under Orpheus and Aglaophemus,
> to be later developed among the Greeks and Italians under
> Pythagoras and finally perfected in Athens under the divine
> Plato . . . The whole world is now in the hands of the
> peripatetics and is divided mainly into two sects, Alexandrists
> and Averroists. Both deny any form of religion . . . But in these
> times it pleases divine Providence to confirm religion in general
> by philosophical authority and reason until, on a day already
> predestined, it will confirm the true religion, as in other times,
> by miracles wrought among all peoples.[10]

Ficino was indeed right about the uniqueness of his historical position. Although the extant works of Aristotle had been almost entirely recovered centuries before, shorn of much of their Neoplatonic additions, and established as the backbone of medieval university education, the complete Plato and full-fledged Christian Platonism arrived as a purely Renaissance phenomenon in the city that is synonymous with the birth of the Renaissance itself, Medician Florence. As can be seen from the previous quote from Ficino, it was in the context of the divine light of Christian Platonism versus the darkness of Aristotelianism that the new philosophy emerged under his hand. Ficino's unique Renaissance Christian Platonism was taught in the Florentine Academy and disseminated throughout the learned West through active personal interchange among scholars. Thus his translations, original writings, teaching and correspondence had wide-ranging influence and affected the visual arts, literature, philosophy, religion and science.[11] Just as this broad influence started as a privately funded, and privately taught, movement in which selected interested individuals took part, so it remained primarily extra-institutional, and was often locked in deadly combat with institutionalized Aristotelianism (this was a constant theme of Robert Fludd's, which was clearly emphasized by Ficino). Though Plato and Platonism came to be studied in the universities, they never enjoyed the primacy Aristotelianism was given in the curriculum. In addition to its mystical-tending metaphysics, Platonism's appeal was based on its broad syncretic quality and the promise of true and certain knowledge beyond the veil, and its ability to provide the possibility of knowledge of the world with geometric and mathematical certainty.

Ficino's great enterprise was added to and given even broader scope by one of his brilliant students, Giovanni Pico, Count of Mirandola (1463–94). Pico was unusual in a number of respects, not the least of which was the astonishing breadth of his learning, acquired over a relatively short period of years. He studied at the universities of Bologna, Ferrara, Padua and Paris as well as at the Florentine Academy. Not only did he become thoroughly familiar with the classical authors available in his time and with the contemporary humanists, he studied all the medieval scholastics as well, and his gift for languages encompassed Arabic and Hebrew on top of the usual Latin and Greek. The advantages of wealth allowed him to build a large and eclectic personal library and devote much time to study, which resulted in a prodigious output in the short life of the precocious nobleman. He became the epitome of the "universal" Renaissance man. Pico's unusually broad range of study convinced him that there was a central core of truth in all that he had read, but that each major philosophical and religious branch had interpreted it in their individual way. He could therefore advocate that all branches of knowledge and religion could be reconciled into a unified whole, and to this end he proposed, in 1486, at the age of twenty-four, to debate 900 (no less!) theses in Rome the following January to prove the point.

Unfortunately for this great enterprise, and Pico personally, a few of the theses ran afoul of some theologians, and Pope Innocent VIII stopped the public disputation, to which all interested scholars had been invited, until the theses could be examined by a special commission. Seven of them were condemned outright as heretical, six were found to be questionable, and Pico's rationale for them was repudiated. When he tried to defend the thirteen heretical and suspect theses, the Pope condemned them all, and Pico fled to France. Arrested there in 1488, he was eventually released to the protection of Lorenzo de' Medici through the influence of the King of France, and spent the rest of his life in Florence writing and in association with the Medici and Florentine Academy circles. It was symbolically significant that Pico died (at age 31) on the same day (17 November, 1494) that the King of France, Charles VIII, who had intervened on Pico's behalf to secure his release from prison, entered Florence with his armies.[12] The invasion by the French and the death of Pico, Lorenzo, Ficino and other leading figures of the time within a few years of each other was a distinct turning-point that marked the end of a great period of intellectual flowering of the Florentine Renaissance.

One of Pico's major contributions to the Renaissance Christian Neoplatonist tradition was the addition of the Hebrew mystical tradition known as Cabala. He saw it as a true and deeper interpretation of the Mosiac laws that God gave to Moses, which could only be passed down orally and in secret to worthy initiates:

> Not only the famous doctors of the Hebrews, but also from among men of our own opinion, Esdras, Hilary and Origen write that Moses on the mount received from God not only the Law, which he left to posterity written down in five books [i.e., the Pentateuch], but also a true and more occult explanation of the Law. It was, moreover, commanded him of God by all means to proclaim the Law to the people, but not to commit the interpretation of the Law to writing or to make it a matter of common knowledge. He himself should reveal it only to Jesu Nave [Joshua], who in his turn should unveil it to other high priests to come after him, in a strict obligation of silence.[13]

Accordingly, thereafter, such initiates as Pythagoras carefully observed this rule, which was confirmed by Dionysius the Areopagite, who stated that the hidden mysteries were passed down by the early religious founders by the spoken word only. And in this way, "when the true interpretation of the Law according to the command of God, divinely handed down to Moses, was revealed, it was called the Cabala."[14] However, problems arose with this tradition because of the various calamities and the scattering of the children of Israel, and the precious wisdom was in danger of becoming lost. Consequently, Esdras was commanded by

God to commit ninety-four books to writing; the first twenty-four were to be made public, and were in fact a reconstituted set of Holy Scriptures; but the last seventy contained esoteric knowledge meant only for the wise, "for in them is the spring of understanding, the fountain of wisdom, and the river of knowledge." These, Pico concludes, are the "books of cabalistic lore," in which he found not only the basic tenets of Christianity, but also echoes of Pythagoras and Plato.[15]

This singular essence of divine truth was passed down through several channels at once: by an esoteric understanding of the Scriptures; through the writings of ancient wise men worthy of initiation into such high mysteries; and through the esoteric knowledge contained in the Cabala. But, of course, these sources did not openly divulge their wisdom, because to do so would "cast pearls before the swine." Even worse than teaching about something that would be beyond the capacity of the ordinary to understand, however, was the fear that the information would be misused by those with only material or evil intent. This was so because with this knowledge comes power – power over natural and supernatural forces at a number of levels, which could be used for good or evil purposes. One could thus become a magus, calling upon either angelic or diabolical powers, according to the ends sought, since there are an equal number and rank of both kinds.[16] As a result, the ancient initiates as well as the later alchemists hid their knowledge in riddles, mysteries, metaphors, symbols and myths, the keys to which were only given to the worthy, who had proven themselves after a long apprenticeship.[17]

A true magus, in Pico's view, is one who studies and understands the divine order of things, and it was to achieve such wisdom that Pythagoras, Plato and others traveled afar. This natural magic seeks to know the harmonious relationships in the universe, and to rightly use them, which is a laudable activity in tune with the divine; to controvert them is evil. "A magus is the servant of nature and not a contriver."[18]

Another of Pico's great legacies was his *Oration*, which was written as an introduction to the 900 theses that subsequently became known as his famous *Oration on the Dignity of Man*. Although it was not published until after his death because of the controversies over the theses, the *Oration* contained two significant features: that man partakes of all levels of creation (as a microcosm), and therefore does not have a predetermined place in the chain of being, but rather rises as high or low as he himself freely wills; and that there is a universal truth which had only been expressed in part by each philosophical school and individual author. On the last point, he says that if one takes the trouble to read all the ancient and medieval authors, as he had, and compares them, then the picture of the real truth can emerge. He could, therefore, boldly propose a harmony between Plato and Aristotle, as Boethius and a few other brave souls had done previously. Philosophizing through the esoteric interpretation of numbers, as practiced from Pythagoras to Plato, is also endorsed by Pico, and in fact it is the main factor why humans are the wisest of animals, in his view.

Pico's mark on the Renaissance Christian Neoplatonist tradition was indelible. Ficino's importance was based on his translation and interpretation of Platonist philosophy, and its dissemination through publication and teaching at the Academy; Pico broadened the scope of this tradition by adding the Cabala and the notion of one central truth behind all philosophy and religion.

Of the sixteenth-century writers in the hermetic-Cabalist vein, a few may be singled out as important for Fludd, though not exclusively so.[19] Johannes Reuchlin's *De virbo mirifico* (1494) advanced a Neoplatonist Cabala with the magical power of the Divine Names as an alternative to Scholasticism, and his *De arte cabalistica* (1517), which utilized Cabalist texts not available to Pico, became the textbook for a Pythagorean-Cabalist mystical number system.[20] In 1525, the first edition of Francesco Giorgi's *De harmonia mundi* was published in Venice; it is a grand scheme of the musically harmonic relationships throughout the universe, which also utilizes new scholarship in the Pythagorean-Platonic-Cabalist train of thought. That Fludd was familiar with and admired this work was shown by Peter J. Ammann, and it was no doubt seminal for Fludd's later theory of universal harmony.[21] Cornelius Agrippa von Nettesheim's *De occulta philosophia* (Antwerp and Paris, 1531; Paris, 1533) became the Renaissance encyclopedia of the occult arts, which promoted Reuchlin's Cabalistic magic at all levels of being. Fludd found this, and other compendia available at the time, quite useful, as indicated by his references to it.[22]

The incorporation of Neoplatonism into a new experimental chemical medicine in the numerous writings of Paracelsus

(Theophrastus Bombastus von Hohenheim) had a major impact on Renaissance science, medicine and philosophy by adding an entire new dimension to the Christian Neoplatonism of the age.[23] That Fludd was an "English Paracelsian" has been well documented, and he refers to Paracelsus as "the well experimented Doctor."[24] The English mathematician and occultist John Dee must also be counted as an important and influential figure of the sixteenth and early seventeenth century, although his influence for Fludd is more difficult to discern, since he never directly refers to Dee or his works. He did, however, own a copy of Dee's *Monas Hieroglyphica* (1564),[25] a thorough numerological-Neoplatonist-Cabalist work, and had access to the remainder of Dee's great Elizabethan library when it was bought by Sir Robert Bruce Cotton. Fludd would have been put off, however, by Dee's angel-summoning with Edward Kelly, and did not approve of continuing to hide the ancient mysteries in complex, undecipherable puzzles as in the *Monas*. In addition, Fludd enjoyed good favor at the courts of James I and Charles I, while Dee did not, and he was unlikely to come to Dee's defense at the expense of his own reputation.[26]

In England, the Platonic revival in fifteenth-century Florence found an early and growing reception. The pioneer English humanists William Grocyn, John Lily, Thomas Linacre and John Colet all studied in Italy, and returned under the influence of the Academy.[27] Colet, who had been in direct contact with Ficino, came back to lecture at Oxford on the Epistles of Paul, in which he made frequent references to Ficino's *Theologica Platonica*, and also used citations from Pico. At Oxford, Colet influenced the young Erasmus, who visited him there.[28] But the greatest Platonist of the early 1500s was Erasmus' lifelong friend, Sir Thomas More, Lord Chancellor under Henry VIII. More was a great admirer of Pico, and translated his biography into English.[29] The *Utopia* was clearly patterned after Plato's *Republic*, and it appears that More's adherence to the Platonist Christianity of Augustine and the Florentines formed the basis for crucial decisions in his life, including his refusal to agree with Henry VIII's break with the Roman Church. Indeed, Erasmus called More's household "another Platonic Academy, (but a Christian one)."[30]

Recent research has also shown that an interest in, and study of, Platonist doctrines stemming from the Florentine Academy and their associated occult subjects were widespread in the English universities and private circles during Fludd's lifetime.[31] As noted in Chapter I, this is where his interest in Renaissance Christian Neoplatonism was cultivated and where he first formulated some of his tracts, which would appear later in the *Utriusque cosmi . . . historia*. Not only was there no official policy of repressing such studies at Cambridge and Oxford, there were numerous questions related to the tradition that were permitted for disputation in the late sixteenth and early seventeenth century. There exist many examples of individuals at the English universities of the time who engaged in Renaissance Platonist and related occult studies, but the most famous remain John Dee and Robert Fludd because of their unabashed advocacy of the tradition in their publications.[32]

Chapter VIII

THE MOSAICALL PHILOSOPHY: FLUDD'S ORIGINAL SYNTHESIS

I. ORIGINS AND OVERVIEW

... in their conceptions my philosophical Works contayne, as it were, a revelation, or a producing of those mysteries of Nature to light, which the ancient Philosophers have hidden from the world's eye by masking and obscuring them with aenigmaticall shadows, parabolical types, and ambiguous riddles.

Fludd, "A Philosophical Key"

As a late Renaissance phenomenon, Robert Fludd's philosophy is distinctly in the tradition stemming from the Medici-sponsored Florentine Academy. Perhaps Pico himself would have attempted a great universal synthesis had he lived beyond the age of thirty-one and not suffered official persecution. But an understanding of Fludd's work only starts with Ficino and Pico, for in addition to his own original philosophical speculations, he had the benefit of a number of ancient authors made more readily available through subsequent humanist scholarship, some influential sixteenth-century writers discussed above, and contemporaries such as Harvey, Gilbert and Ridley.

Most of all, however, the numerous authors and Biblical quotations cited in his works are employed to build an intellectual framework for his own spiritual perceptions. He writes as one who spent many years scientifically defining the context for his spiritually experienced universe. All phenomena are related to the central truth of the divinely created universe and its "great chain of being," from the highest realm to the lowest. Following the lead of Pico and others, as well as his own sense of the non-material, he attempts to reconcile his perception with the great Middle Eastern and Western mystical traditions, both philosophical and religious, in a rational, scientific way. The process of his thinking, and therefore the structure of his universe, is that of art; the content he attempted to make scientific.

His universe was created according to musically harmonic relationships, following Pythagoras, Plato and others, with the proportion of pure spirit diminishing directly as the descent is made from the most exalted level to the most material; the dynamic of its action is the cyclical, alternating effect of the outpouring and indrawing of the divine essence or holy spirit. All of Fludd's arts, science and medicine were founded on this scheme, and his argumentation normally proceeds, as a natural consequence, from the spiritual to the material. Thus also for him a science that generalized from "mere" physical phenomena and mathematical formulations derived from them is based on the least spiritual part of the universe, and therefore is the least true or important, and, what is worse, may even be diabolical. For example, he admonishes an imaginary "calumniator" in true esoteric Christian Neoplatonic fashion: "Why is dull sense thy only guide, the opiniative imagination the blind and dim lamp which giveth thee light, leaving thy divine essence wholly void of action, in which only consisteth the real vision of God and Verity?"[1]

Although Fludd followed in the tradition of the Florentine Academy and the subsequent authors mentioned above, his uniqueness was in integrating all of the previous work into one grand systematic synthesis according to his particular arrangement. The tradition and the major works he drew on are clear, but his comprehensive and extraordinarily elaborate construction was formulated from his own speculations about the nature of the Neoplatonic universe, and was derived in large part from his alchemical experiments with wheat. Perhaps the most outstanding feature of Fludd's works is the outstanding quality, and number, of fine copperplate illustrations that illustrate them.[2]

Taken together, Fludd's collected published and unpublished writings (discussed in more detail in Chapter V), which run the gamut from metaphysics to medicine, creation to the arts and sciences, and contain a myriad of fascinating and beautifully executed plates, the quality and quantity of which are not to be found in any other set of philosophical works, are a notable achievement for the age. It was not, however, an achievement which won its author undying fame and lasting honor, as we have seen.

Above all, the foundation of Fludd's philosophy was religious. It was built upon the wisdom of Moses, which he drew from his interpretation of the Bible (including the Apocrypha) along the way prepared by Ficino and Pico, which therefore included the mainstays, Plato and Hermes Trismegistus. All other authors quoted by Fludd in support of his arguments (of which there are many, taken from encyclopedic compendia as well as individual works) simply amplified or corroborated the basic tenets of these three sources. Fludd explains his use of the Bible as the basis for his philosophy in the preface of the *Mosaicall Philosophy*:

> My desire is (Judicious and Learned Reader) that it may not prove offensive to any, if (in the imitation of my Physicall and Theo-philosophical Patron St Luke) I mention and cite the testimony of Holy-Writ, to prove and maintain the true and essentiall Philosophy, with the virtuous properties of that eternal Wisdom, which is the Foundation and Corner-stone, whereon it is grounded . . . If God therefore in and by his Eternal Word or Divine Wisdom, hath first made the creatures, and sustained the same unto his present; How can a reall Philosopher enucleate the mysteries of the Creator in the creature, or judiciously behold or express the creature in the Creator . . . but by such rules or directions as the onely storehouse of Wisdom, namely the holy Scriptures have registered, and the finger of that sacred Spirit indited for our instructions?[3]

He insists, however, that in no way does he wish to encroach upon or alter "those usuall Tenets and Authentick rules in Divinity which have been long since decreed and ordained by the Ancient Fathers of the Church."[4] On the other hand, there may be times when he may *appear* to present unorthodox doctrine, because

> . . . it is certain, that one and the self-same place in Scriptures hath a two-fold meaning, to wit, an internall or spirituall, and an externall or literall; and either of these two senses are true and certain, though they seem to vary or differ by a diverse respect . . . [5]

In all instances, therefore, Fludd considers his arguments based on Scripture to be the epitome of Christian orthodoxy.

The reason one could rely on the writings of Plato and his followers, as well as those of Hermes Trismegistus – Fludd cites them almost always as having equal weight with Holy Scriptures – is this:

... it appeareth by his works, that *Plato* had Knowledge of the Word, and had read the Books of *Moses*; and for that reason he was called, *Divinus Plato, the divine Plato*. In like manner, the excellent Philosopher *Hermes*, otherwise termed *Mercurius Trismegistus*, expresseth plainly that he was not onely acquainted with *Moses* his books, but also was made partaker of his mysticall and secret practice, as by his Sermons, which he calleth *Pymander*, a man may plainly discern, where he doth mention the three Persons in Trinity, and sheweth the manner of the worlds creation, with the elements thereof, by the Word. And therefore of all other ancient Philosophers, I may justly ascribe divinity unto these two.[6]

Showing how truly he adhered to the syncretism of Pico, Fludd says that the ancient philosophers of note gathered their knowledge from many sources and centers of wisdom:

... it is most apparent, that some of the Greekish and AEgyptian Philosophers, namely, *Plato, Pythagoras, Socrates, Hermes*, etc. did so instruct their understandings, partly by the observation of their predecessors doctrine [i.e., that of Moses], and partly through the experience, which in their long travails and peregrinations they had gathered, among the AEthiopians, AEgyptians, Hebrews, Armenians, Arabians, Babylonians, and Indians, (for, over all or most of these Countries did *Plato, Pythagoras, Hippocrates*, and others of them travell, for the augmentation and increase of their knowledge) ... [7]

One other ancient author also appeared often on Fludd's pages, primarily as the Devil's advocate: Plato's most famous pupil, Aristotle. For Fludd, what little of merit there is in Aristotle's writings was stolen from his master, unacknowledged, "masked under strange titles," then presented as his own doctrine. What makes Aristotle particularly reprehensible is that even though he was exposed to the true wisdom through his teacher, he remained "alltogether ignorant of the central truth thereof," and so he proceeded to build "a bastard Philosophy which did differ in shape and essence from the true Foundation."[8] The serious consequence is that Aristotle's first principles are close enough to the true ones to seduce seekers of knowledge into further study, but this leads them down the path to a spurious and worldly philosophy, thus away from the true wisdom of Christ. One is warned, "It well behoveth therefore each Christian, to be wary in his reading the Ethnick Philosophy, and to consider seriously before he wade too far in it . . . "[9] Fludd feels it is essential to make the point that

> . . . the Philosophy of the Ethnicks is false and erroneous, both in regard it is founded upon the wisdom of this world, which as St *Paul* teacheth us, is but meer folishnesse in the eyes of God [in margin: I Cor. 3.19] and then because it contradicteth the truth, and consequently is not issuing from the Father of Light, which is in Heaven, but from the Prince of darknesse, who reigneth beneath, wherefore this kind of wisdom, or σοφία is termed by the Apostle *James*, *Terrene, animal, and diabolicall*. [James 3.5][10]

What horrifies Fludd is the realization that "even in this later age," Satan, the Prince of Darkness, still has the upper hand in the struggle with the Forces of Light, since

> . . . this Terrene Wisdom or vaine Philosophy, which is dawbed over with dark ignorance, hath the dominion or upper hand, and so by that means Christ, which is the true Wisdome, is daily crucified among some Christian Philosophers, and buried in darkness, through the mysty and ambiguous clouds of that cavilling, brabling, heathenish Philosophy, which they so adore and follow, with their Master *Aristotle*, as if he were another Jesus rained down from heaven, to open unto mankind the treasures of the true wisdome.[11]

Since education is still mainly in the grip of Aristotelian concepts, it is still in dire need of reform to bring it back in line with true Mosaical Christian wisdom.

The immediately preceding quotations fairly and accurately show the basis for Fludd's unified concept of theology, philosophy, science and medicine, no matter how elaborate or complicated in detail they become in his numerous works. Fludd was a devoutly Christian Neoplatonist, or, as he would call himself, a Mosaical Philosopher, since he begins with what he believes to be infallible principles delivered from God directly to Moses and his followers. Since he made some unique contributions to his grand synthesis of knowledge, over and above what was formulated originally by the Renaissance Florentine Neoplatonists, he felt justified in calling his philosophy "Fluddean" as well as Mosaical.

104

The most striking original feature of Fludd's works, as noted above, is his explicit, detailed and highly organized Neoplatonist-Cabalist-alchemical interpretation of the universe in both his descriptive (and often sermonizing) philosophy and his diagrams. The often lavish and elaborate copperplate illustrations by de Bry and others not only served to clarify his concepts, but also have been more responsible than anything else for keeping his name from passing into total oblivion. Indeed, these illustrations have provided the basis for some of the modern revival of interest in the English physician. They are often found reprinted in a variety of modern books dealing with the Renaissance (including, for example, the works of Shakespeare), and they were responsible for the rarity and high price of Fludd's books well over a century ago.

In addition to his unusually detailed diagrammatic schema of the corresponding microcosmic and macrocosmic worlds, Fludd added to his syncretic philosophy some distinctly new concepts which he thought were implied in the Mosaic books (augmented by Plato and Hermes). The most important among these new concepts (all of which are discussed in more detail below) are the following. First, he concludes from Genesis that the common or secondary elements (earth, air, fire and water) ultimately derive from the primary (or primeval) element water. This, he thinks, is a necessary conclusion from the story of creation. Second, he viewed the sun as the intermediary body between the heaven and earth, containing the spirit of God. The sun in the heavens is exactly analogous to the heart in man, the Microcosm.[12] Third, he invented the diagrammatic representation of the emanation of spirit downward and matter upward by two interpenetrating pyramids. Of all his discoveries "by contemplation," Fludd was most proud of this one; many other authors had described or diagrammed in one form or another many of Fludd's representations, but the pyramids were his own innovation. Last but not least was Fludd's attempt to effect a grand pictorial synthesis of the harmonic relationships of the universe in his *Monochordum Mundi*, which coordinates the emanation of the forces of creation in a series of octaves according to Plato's heptachord, Proclus' division, the Cabalistic interpretation of the divine name, the numbers of its geometric progression, the four elements, the three heavens and the angelic hierarchies.[13]

For an understanding of the framework of Fludd's Mosaicall

philosophy, one must start with his notion of creation and the progressive development of the macrocosmical universe along Pythagorean-Neoplatonist lines. It is from this construction that he derives his theory behind man, the Microcosm, and all the diverse topics covered in his various published works, which range from the occult arts to optics, and from surveying to anatomy and medicine.

II. THE MACROCOSM

In several works, besides the great *History of the Macrocosm*, Fludd tells his version of the process of creation. Though they may vary slightly in details given, each is the same in essence.

Within the Eternal Archetypal Unity of God (YHVH), there exists a two-fold principle, which appears as polar opposites completely contrary in properties. They correspond to the Nolunty and Volunty of God: the first of these is defined as remaining in potentiality, or not willing (nilling), reserving itself within itself, and is expressed by darkness and privation (Dark Aleph); and the second is the willing, or acting of God, represented by the outpouring of life-giving, and sustaining, light (Light Aleph). The consequence of both of these two opposite principles being in the essential nature of the subsequent created universe is that there are two main branches of existence: the dark side brings discord, evil, cold, congelation, rest, death, privation, negation; but the light brings concord, goodness, heat, resolution, motion, life and position. However, unlike as in Manicheanism, these two contrary principles both remain aspects of God's unity, to which they will return at the end of days, and the great mystery of why this was so will be revealed when the Seventh Seal is broken.[14]

In Fludd's version of the beginning of the universe, he posits before the act of creation the existence of an uncreated *materia prima* or first matter, which he also calls the Hyle (Ain):

> This Hyle is meere nothing or puissance to be somethinge, againe we find that a thing in puissance or *posse* is farr different from that which is act or *esse*, wherupon it foloweth that being the first matter . . . it is absolutely nothing in act; now since all creation is the reducing of that which was never really before into act (for the presence of God made all things to have forme

which before were nothinge) it is to be considered consequently that creation is nothing else but an inactuation or reducing of nothing in deed into something . . . [15]

Fludd further describes the Hyle as an "infinite mass or darke fog, as black as pitch, without any consideration of the least sparke of light within it."[16] He admits that the postulation of this uncreated Hyle brought criticism from some who claimed that he was espousing erroneous religious opinion by maintaining an entity to be coeternal with God. He denies any conflict with Christian teaching, bolstering his argument with this testimony: "Plato could liken this matter to nothinge else but a vision in a dreame, which waking playnly appeareth fantasy, and St Augustine compared it to silence, which in respect of speech was mear nothinge."[17]

The action of the first outpouring of divine light transforms this "voyd and inane darkness" by fecundation into a primeval Chaos, which is dark and deformed in its exterior, but contains the "actual" elements within it:

. . . after the aparition of God out of darkness the viscous spirits included in darkness, having embraced an infinite company of sparks or beames of light which penetrated into the dark abyss, turned immediately that mass of Hyle or first matter into a Chaos, which is a rude and indigested matter in whos belly the five elements wer so irregularly included that they jarred . . . and contended with one another impetuously.[18]

The five elements referred to here are (in Fludd's order of "visible grossness to visible subtilty"): earth, water, air, fire and quintessence.

This first appearance of the divine light took place on the first day of creation. On the second day, the primeval Chaos, which Fludd likens to the waters of Genesis, which were without form and were "like a troubled fume or vapour," is acted upon by the divine spirit (Elohim Ruach): " . . . the motion of the spirit uppon the waters [proceeded] from the word Fiat with the [decline] of darkness towards the center with the information of the waters, which were the vehicles of that operating light . . . "[19]

The outpouring of divine light, pushing darkness downward to a region of maximum density, was completed on the third day. In this way, according to Fludd, the three heavens were created:

" ... the highest heaven (the Fiery or Intellectual Heaven, the Caelum Empyreum) was made the first day, and the middle region or starry heaven (the Sphere of Equality, the Platonic Sphere of the Soul, the Aethereal Heaven, the Caelum Aethereum) the second day, and the third day the lower or elementary heaven (the Caelum Elementum)." This is also the same act of the separation of the waters: the Empyreal realm was made of the brighter, worthier waters; the Elementary world received the grosser, darker and viler waters; and the intermediate Aethereal heaven was equally divided between the extremes[20] (Plates 2–5).

Following Plato's account of creation in the *Timaeus*, Fludd asserts that "the degrees of darkness ... did descend in measure and proportion towards the centre of the world ... "[21] The downward pressure of the divine light caused the darkness to become double in the middle region on the second day, and treble on the third in the lower region. The continuing action of the informing spiritual light in the elementary world brought forth the three earthly kingdoms: "Hence ... came that Chaos which contayned the three elements namely the sphere of fire, humidity and earth. And hence proceeds the triple kingdome of nature in composition, namely, Animal, Vegetable and Mineral ... "[22]

On the fourth day of creation, after the earth's position was made fast, Fludd (now following "Moses, the Father of the Theosophists") insists that it was God's will that after the natural position of things were fixed, there should be a natural God: " ... the creatures of the first heaven wer made the first day, and the creatures of the lowest heaven wer made the third day, as minerals and vegetables ... [and] the creatures ... of the middle heavens wer not made till the fourth day, namely the sonne, moone and other starrs ... "[23]

This natural God, the sun, is the seat of the fifth element, the quintessence (also termed condensed spirit: it served as the vehicle for the soul's descent into matter), which was extracted on the fourth day from the earth. It was placed in the middle of the middle heaven, where it participates equally in the spiritual and material. Its beams vivify the world temporally, causing multiplication in the sublunary compounds.[24] In its central position, the sun is related to both God above and the earth below in accordance with the harmonic intervals of music (see Plate 3): " ... this spirituall and supernaturall God, the invisible son of God the father (as the World's sonne is called of Plato the visible sonne of God) governeth from the sonne upwards by a spirituall and invisible Diapason, as the visible sonne operateth downward by a visible and materiall one."[25]

Fludd's favorite invention to demonstrate the relationships between the divine light and dark matter was the inverted pyramids, as in Plate 2. The "formal" pyramid has its base touching the presence of God and gets thinner as it goes downward to earth; the earth is the fountain of matter and passion, whose representative pyramid diminishes as it reaches upward, ending in a point at the top of the Fiery or Empyreal Heaven. The two pyramids are exactly equal in the center of the Middle of Aethereal Heaven, and there God has placed His visible representative, the sun[26] (Plates 2, 3, 5, 7, 8).

A vital aspect of Fludd's universe was the idea that God is "all in all" (I. Cor. 15:28), that is, the divine spirit not only brought about the created universe, but continues to bring sustenance and multiplication at all levels of being. This vivifying catholic spiritual or angelic presence is called the *Anima Mundi* or Soul of the World by the Platonists, Metatron by the Cabalists, and is also called the *Donum Dei*, the Gift of God, *mens divina* or the emanation of the Word. It is that which provides the Image and Similitude of the Divine Wisdom.[27] The other major feature, discussed further below, was that in the entire chain of being, descending from the Trinity above, through the nine orders of angels in the empyrean region, the planets, moon and sun in the middle realm, and the fire, air, water and earth in the lowest sphere, each part is related to the other as the major harmonies of a monochord: the double octave (4:1), octave (2:1), fifth (3:2), and fourth (4:3).[28] Man, the Microcosm, is structured in exactly the same analogous way, and all of this is reconciled with the Cabala. Thus we have the interpenetrating pyramids of spirituality and materiality, the three regions, the harmonic arrangement of the constituent levels, the exact parallels between the Macrocosm and Microcosm, and the Cabalist equivalents, all of which form the basic structure of Fludd's universe.

The spiritual light–dark dichotomy, and its manifestations as heat and cold in the material world, form a central part of Fludd's philosophy, science and medicine when combined with the root or primeval element water. In the beginning, creation proceeded by the progression darkness-water-light; this original threefold mani-festation Fludd took to be a reflection of the Christian Trinity[29] (Plate 6). Since all the lower or secondary elements ultimately derived from the primeval water (and could be converted into one another because of their common root), it is the action of heat or cold on them which causes change for good or evil. The heat or cold is a manifestation of the outbreathing or inbreathing of God, the outpouring or indrawing of the divine will, divine light or divine spirit; or, as Fludd sometimes puts it, the nolunty or volunty of God.

From this necessarily brief description of Fludd's Macrocosm, let us turn to the Microcosmical world of man and see how Fludd shows these divine truths by his experiments.

III. THE MICROCOSM

In his unpublished manuscript "The Philosophical Key," Fludd provided his readers with a literary version of the origins of man which coordinates with his mystical and diagrammatic schema. It also clearly expresses man's relationship to the greater cosmos, and how man begins to be aware of this relationship. This version combines some of Plato's *Timaeus*, classical mythology and other sources. Addressing his detractors, Fludd askes them to envision the following:

> Think, that in the first place thou wert clay or dust, or earth, or ashes . . . And then that Demogorgon[30] (burnished all over with sacred and eternal fire, drawing the dim courtaynes or dark tapistry of his high iluminated Palace, attended by Aeternity and Chaos, his two obedient and loyall vasals before the beginning) did vouchsafe with lowde Ma[ty] to respect and looke on from above thy sensles mould; and for thy sake did command Chaos to discharge by abortion her troubled wombe of Litigium,[31] that after her purification she might conceave and bring forth a second frute of a more worthy condition, that is to say, Pan or Universall Nature.[32]

Then, Demogorgon, the creator god, incites Chronos or Time, the first born son of Aeternity, to join with "that fayer and humid impe" Pan, in order to "direct him [Chronos] in his diurnall and naturall course . . . that so by both their ayds and assistances

110

together thou (ÔMan) and all things else necessary for thy preservation might be produced out of darkness and created of meere slime, and, as it wer, nothing."[33]

Now it was deemed necessary by Demogorgon for this senseless mass of clay to have the ability to know its divine origins:

> Lastly apprehended in thy benumed sence, that thou beholdest this most high and mighty President of light, to impart unto thos his first universall creatures and ministers a portion of his divine fire, which he breathed from his owne nostrils to be infused into this deformed mass and sensles mould of thine to illuminate it.[34]

This descent of the divine fire through the previously created hierarchies proceeds according to harmonic proportions, as shown in Plate 5. Once the infusion of divine illumination takes place,

> . . . thou immediatly therupon wert of vile and senceless clay raysed up a living, sensible, and reasonable creature, resembling the very image of that heavenly beam of light, by which thou breathest and invest this angelicall forme and habit belonging unto man . . . [35]

At the moment of inspiration man was afforded a last fleeting glimpse of the Creator in order to know His majesty:

> Then behould at the very instance of this thy perfection the goulden type of thy Creator (inthroned on the highest clouds, wafting upon the swift wings of the winds) who being accompanied with the many legions of angels (circling Him all about with melodious hymnes) did draw the dark courtaynes of his clowdy pavileon over the splendour of his presence, and so vanished quight out of sight, leaving thee to the guiding and preservation of Nature and Time.[36]

Fludd then has Nature and Time, personified as Pan and Chronos, deliver the following lecture to man:

> Ô Anthropos . . . how infinitly art thou obliged to the greatest Monarch Demogorgon, who by infusion of his sacred and never fayling fire into thy senses and dead nostrels hath framed thee after his owne image! and for thy better safty in the accomplishment of a dangerous pilgrimage which thou must enterprise in this wide and slippery world, he hath ordayned us,

as a double wach or garde unto thy person, to defend thee from the malice of Litigium, that spurious and abortive sonne to foule obscurity.[37]

Pan continues the narrative alone, explaining that it was Demogorgon who

... hath blowen up every on of my members with Musical sympathy engendering in them this proportion, measure, and melodious harmony, so that I receave my rithmical and metrical action from him and every wel agreeing consonance from the course of this my fellow, and his servant Time, who dewly ... leadeth me about the ecliptick line of his methodical and most regular motion.[38]

Pan says that she hourly blesses her celestial father, who has made her of divine symmetry and to represent

... the universall mass of watry spirit, which he hath sublimed and refined by the rectifying fire of his heavenly Alchemy ... whereby I have the glorious image of him self; for through the bright characters which he hath impressed into my perspicuous and cleare substance, he hath made me partaker of the draps of his unvalued essence, which doe contayne the bright reflexion of his never-fading presence.[39]

After being extracted from her mother by Demogorgon, Pan explains that

... then was I commanded in they behalf to doe as followeth. Ther is (quoth Demogorgon, speaking in his highest Ma[ty]) a hidden mystery, which I wil make playne and manifest unto the world by mine omnipotent vertue. And this it is. I have decreed and established in my secret councells to create a lesser world and ordayne him sonne to my self, and thee (Ô Pan); for thou shalt be his mother, and I wil be his father: Him therefore will I exalt to high estate and dignity and make him thy companion.[40]

For the inner spiritual structure of man, Demogorgon provided that

... I will replenish him with a supernaturall splendour floting and swimming in the bright streames of the Empyreall and Aethereall spirit, by the vertue of whose action and inclination he shall first of all from the higher and most refined spirit, as immediately ishewing from my presence, be replenished with discretion; and then from thy holsome brests and charitable bosome, even the sonny spheare of thy soules equality [i.e., the sun] shall he continually suck in his breath, and attracting with it by a magneticall sympathy the milky vertue from thy [marginal note: that is, from the sonne & moone] glittering and goulden duggs, he shall harbor and possess a vital soule.[41]

For the "entertaynment and preservation" of the vital soul, Pan was directed to make a corporal temple or palace to be forged out of her grosser elements. This temple is to have seven openings to receive sense impressions (i.e., two eyes, ears and nostrils and one for taste) which correspond to the seven planets ("the mayne windows or casements of heaven"), and a universal sense of touch. Demogorgon further decreed to Pan:

. . . let him [Man] . . . in every respect imitat and resemble thyself, who fillest the greater world with thy presence, that ye may be both the true patternes and examples of my self.[42]

For this purpose, Demogorgon further decrees that

. . . as my cheef dwelling is in the highest top of the purest heavens, so also shall that intellectuall spark of my never-fading spiracle inhabit the most eminent and capitall region of this small world.
    Then in his midle spheare shalt thou erect a pavillion called the hart, which, lik the sonne in the greater world, shall send forth his essentiall beames circularly from his centre, that thereby they may animat and vivify every member of this so well erected a Microcosme.[43]

The Creator calls Chronos to duty at this point; just as he carries out his office of supervision of the heavenly motions,

. . . he shall justly guid and proportionat the minuts of his lif and dayes, observing carefully that the motion of his pulses be obedient to just measure and harmonicall proportion, and that their Systole and Diastole doe live together in peace and concord . . . [44] (See Plate 17).

After a description of some of the workings of the internal organs in the middle sphere, the lowest part is fashioned:

Finally his gutts and entrailes (being appropriated to receave all

113

draggy and terrestriall excrements) shall represent in him the earth of the greate world. His urin shal have relation to the salt and brinish seas. His stomacks kily substance shall respect the sweat fountaynes of water, which issue from the bowells of the earth and are nourished by the clowds of heaven. His blood shal in nature imitat the aery element; and the fiery circle of this litle world shal intersect and pass through the vessell of the Gale; for by it shall that cholerick receptacle be nourished. And lastly his spleen must participat of the earthly disposition, for as much as it is the receptacle of the thinner mucosity and slime thereof.[45]

And thus, says a well-pleased Demogorgon, "shall ye mingle and put together the greater world's thinner heavens with his grosser elements to accomplish in perfection this well framed tabernacle of the lesser world."[46]

The correspondences of the Microcosm and Macrocosm as detailed above are graphically depicted by Fludd in several illustrations. Plate 2 shows the basic relationship of the three heavens and the inverted pyramids previously mentioned. Plate 8 again shows the correspondence of the three heavens, the heart and sun in the middle sphere, and the lower region. This diagram relates in the lower region fire with the choleric, air with the sanguine, water with phlegm, and earth with the excrement.

The uniqueness of man in Fludd's schema reminds one of Pico della Mirandola's *Oration on the Dignity of Man*. Although the two are not the same word for word, there are definite parallels. We can compare, for example, where Pico says the following:

God the Father, the supreme Architect, had already built this cosmic home we behold, the most sacred temple of His godhead, by the laws of his mysterious wisdom . . . But when finished, the Craftsman kept wishing that there were someone to ponder the plan of so great work, to love its beauty, and to wonder at its vastness.

Therefore, when everything was done (as Moses and Timaeus bear witness), He finally took thought concerning the creation of man. But there was not among His archetypes that from which he could fashion a new offspring . . .

All was now complete; all things had been assigned to the highest, the middle, and the lowest orders . . . At last the best of artisans ordained that the creature . . . should have joint possession of whatever had been peculiar to each of the different kinds of being . . . [47]

In the above quotation Pico refers to Moses and Timaeus (i.e., Plato), and elsewhere in the *Oration* he specifically refers to Hermes Trismegistus. These three are Fludd's mainstays. Returning for a comparison to the "Philosophical Key," we find that during Pan's address to man she says,

> Ô Microcosmos, the sonne of Demogorgon in respect of thy divine and formall fire, and the cheefest offspringe and image of my self considering thy materials, didst prove suddenly (on being infused with the divine nectar) a living creature indued with sence, discretion and high wisdome, in so much that thou wert the prince and most especiall composition of all that was created. [48]

For both Pico and Fludd, man should try to understand his divine origins and seek the highest good, which is a return to the divine source within. Pico's notion was that man would transform himself into whichever part of his nature he cultivated. If a man cultivates the vegetative, he becomes like a plant, and brutish if he dwells in the sensual part. But,

> If rational, he will grow into a heavenly being. If intellectual, he will be an angel and the son of God. And if, happy in the lot of no created thing, he withdraws into the center of his own unity, his spirit, made one with God [he] . . . shall surpass them all. [49]

Fludd, basically following the same Neoplatonist sources as Pico, insists that man should follow the same path:

> Retourne, Retourne (I say) unto thy self, subject thy body unto thy reasonable soule by diving into thy inward treasure, and then prostrat and submit that thy mental and spiritual part unto thy God; for so shalt thou be made on spirit with him . . . by which thou shalt be glorified and exalted. [50]

One must be wary of worldly distractions,

> . . . for thos ar the slights of Litigium by making charmes to divert thyn eyes from the glorious face of Demogorgon, after whos vision the soule so thirsteth, as wishing her highest good and happines. Wherfore I cordialy admonish thee to ascende

from this world unto God, that is to penetrat quite through
thyselfe: for to clime up unto God is to enter into thy self . . . for
this is the quiet rest and repose of our harts, namely to be fixed by
a strong desier unto the love of God.[51]

The Neoplatonic quality of the two preceding quotations is
unmistakable. Man contains God within, to whom the soul desires
to return if not obstructed by the diabolical character of worldly
charms.

Fludd believed very strongly in these precepts. All the evidence
indicates that his personal life was one of piety and moderation. As
indicated earlier, Fludd stated in the "Declaratio Brevis" that at
the age of about forty-five he was still perfectly chaste, and in the
"Philosophical Key" he dismissed those who have accused him of
incontinence by assuming that they judge him by their own life.[52]
He sums up his views as follows:

> Wherfore nothing can be more notable and excellent unto a
> blessed lif, than [sic] by closing up the carnal senses, which ar
> without the flesh and the world, to convert every outward
> affection into the inward self, and to referre each forrene and
> alienated inclination and intention from the lustful appetits of
> mortal men unto one self, and by that means only to conferre, as
> it were, and talke with Demogorgon, who dwelleth within *thee*.[53]

A corollary of this idea that all men are created with the divine
center is the brotherhood of mankind:

> Thus I say, wilt thou know, that each man is thy brother, and
> that thy brother is a part of thy self, and al men ar but on, and
> the self same thing in specie, which is in effect the very unity or
> act, and essence of God himself masked from the sight of
> unworthy men with the material mantle of Nature.[54]

Pico asks the question, "But by what means is one able either to
judge or to love things unknown?"[55] Put another way, how can we
come to know our divine origins? The answer for Pico and Fludd is
the same: we can be apprised of the true wisdom through certain
sacred texts and through an investigation of nature, wherein the
true character of things revealed in the texts can be confirmed. Of
the sacred texts, we can learn from Moses and the other ancient
fathers who were initiated into the divine mysteries, including

pagan, Hebrew and Christian sources. Each of these civilizations produced bodies of writings which were meant for the masses, but there was a corresponding set of secret teachings for each of them which were only to be revealed to those worthy of receiving such divine truth. Of the pagans, Plato, the Neoplatonists and Hermes Trismegistus are the principal sources of truth, because they were initiates. The Cabala contains the central truth of the Hebrew texts, and the secret Christian teachings were revealed by Pseudo-Dionysius, thought to be St Paul's disciple. That the true wisdom had been known to the ancients and was the same as Christian teaching was confirmed by some of the Church Fathers, principally St Augustine:

> The thing itself, which is now called the Christian religion, was with the ancients, and it was with the human race from its beginning to the time when Christ appeared in the flesh: from when on the true religion, which already existed, began to be called Christian.[56]

To Renaissance occultists, the above sources were all thought to have widely different origins, but were similar because they represented different approaches to the one true wisdom. The actual reason they were alike is because of their shared Platonic or Neoplatonic influences.

The other way to know the divine truths was to extract the hidden mysteries of Nature, God's handiwork. Here again, Fludd and Pico were in basic agreement, except that Fludd, a practicing alchemist, realized his shortcomings and shrank from the dangers of high magic.

In his *Oration*, Pico distinguished between the neocromancer or black magician, who uses demons for his work, and the magus, whose work "is nothing else than the utter perfection of natural philosophy."[57] He cites Alkindi, Roger Bacon and William of Auvergne as examples of true magi, and notes that Plotinus called a magus the servant of nature, not a contriver. For Pico, a true magus,

> . . . abounding in the loftiest mysteries, embraces the deepest contemplation of the most secret things, and at last the knowlege of all nature. [The magus], in calling forth into the light as if from their hiding-places the powers scattered and sown

in the world by the loving-kindness of God, does not so much work wonders as diligently serve a wonder-working nature.

[The magus], having more searchingly examined into the harmony of the universe, which the Greeks with greater significance call *sympátheia*, and having clearly perceived the reciprocal affinity of natures, and applying to each single thing the suitable and peculiar inducements . . . brings forth into the open the miracles concealed in the recesses of the world, in the depths of nature, and in the storehouses and mysteries of God, just as if she herself were their maker . . . [58]

Pico insists that the efforts of the magus are divine and salutary, inasmuch as

. . . nothing moves one to religion and to the worship of God more than the diligent contemplation of the wonders of God; if we have thoroughly examined them by this natural magic we are considering, we shall be compelled to sing, more ardently inspired to the worship and love of the Creator . . . [59]

This concept of using natural magic to investigate the mysteries of the unified, sympathetic harmony of the universe precisely describes Fludd's approach to exploring natural phenomena. It provides the basis for his alchemical experiments, as well as for his belief in the efficacy of the weapon-salve cure.[60]

Although Fludd agrees with Pico's ideal of the possible accomplishments that could be made by a skilled magus, the English doctor found his own abilities limited to observations made from some relatively simple alchemical experiments. His main experiment was with wheat, which he said provided him with the basis for his entire philosophy (when reconciled with the Mosaic and Neoplatonic texts).[61] That his conclusions from his alchemical studies of wheat warranted building an entire philosophy on them is once again justified by Fludd's concept of the universe (as expressed by Pico above):

. . . this most perless Queen [Universal Nature] sitting most abstrusly in her centrall pallace [the Sun] feeds the composition of each of hir three kingdoms, streaming forth the essence of hir beams from the middle pointe unto the very skirts and margins thereof.[62]

These beams, streaming out into the animal, vegetable and mineral kingdoms, form the bond which holds the four elements of each kingdom in harmony:

This is . . . the invisible fire of Zoroastes & Heraclitus, the essentiall Ligament of the Elements & that vertue of true comixtion, wch causeth so compleat a union amongst the foure dissonant natures of every one of the foresaid kingdomes in the sublunary world.[63]

Although the things of this world differ in outer appearance, they are exactly the same in essence:

This is she [Universal Nature] who imprinteth the convenient character and forme of every creature in his proper kind, whereby on thinge is distinguished by an essentiall difference from an other . . . and yet is she not many Natures but indivisible and only one in number, governing like the very image of the generall Creatour in all & over all, for as much as she being the life of all things is with the individuall soule of the universe sayd to be in all & therefore in every part of the Macrocosmicall edifice . . . [64]

Fludd then maintains that

. . . she hath chosen for her cheefest mansion in the Animall Kingdome the body of Man, that most excellent of all sensible creatures; in her Vegitable Empire she hath elected Wheat that most worthy of all vegetables for her richest tabernacle . . . in her Minerall nature she most delighteth in & principaly inhabiteth her goulden palace [i.e., gold], burnishing it about with the streams of her brightest glory.[65]

It would be a most admirable feat, he says, for the adept or "Artist" to be able to summon the fire of celestial alchemy in order to be able to recreate and thus understand the inner and outer structure of the animal kingdom. By so doing, the Artist would,

. . . by survaying of his most secret and hidden regions . . . quickly be taught in a true vision, to know himselfe & to discerne the highest heaven of his inward Man, that is to say the most intricat & central closet of his hidden spirit, in wch the Maiestical presence of that intellectuall beame hath his

119

residence, by whos sacred presence he excelleth & hath command over the living creatures in this world.[66]

But Fludd freely confesses that he has never achieved the skill of such an adept, and also admits to a fear of dealing in this kind of high magic:

> And verily although I professe myselfe to be ignorant of that excellent skill eyther to understand rightly thos hidden mysterys or the regiment of that aethereall fyre by wch such glorious effects in Nature may be accomplished, & therefore am debarred and denied from entring into that straight path wch conducteth unto the vision & fruition of bliss: although (I say) the sharpe punishment of Prometheus hath added terroir unto my thoughts & deprived me of the hardness to steale any of this excellent fire from heaven, neyther hath mine ignorance & unworthines permitted me to obtayne it from above through grace; and piety, despayre & dread of Joves displeasure commandeth me to forbeare violently & vainely to attempt the winninge & pulling of it by force from the adamantine skyes, least by the dent of a thunderbould I should wth the ambitious & audatious Typhon be buried in the bowels of the darke earth & with restless flames remayne tormented for my rebellion . . . [67]

This statement is important for understanding Fludd's viewpoint. Whereas he thought it was possible for an "operatour" to achieve the wonder-working of a high Renaissance magus as described above, he himself felt unable to attain this ability and had a great fear of the divine wrath for even attempting to gain it. Thus while he believed, as Ficino and Pico did, in the possibilities of high magic and the scheme of the universe that went with it, Fludd never involved himself with any kind of magical diagrams, ceremonies, incantations, purifications or the like. His magic was limited to the sympathetic natural magic of medical cures, discussed earlier.

Since he was unable to perform the feats of a great magus, and since the structure of nature was the same in each of the three kingdoms, Fludd had to content himself with rather simple alchemical experiments on nature's chief mansion in the vegetable kingdom, wheat:

... wth the common & spurious Alkimist, that toyish ape, and superfluous imitatour of Nature, I will heare presume to dive so far into the hidden parts, drowned as it were in the darke composition of this rich and unvallued vegetable, hid from the vulgar eye, as the smal sparke of that super-celestiall fire wch my Creatour hath infused into me shall direct, being contended with the lott and parcell of that curious experience wch mine owne labours & industrious search into the secret natures of things have taught me . . . [68]

But, Fludd contends, the rewards of delving into nature's mysteries are great:

... what happines & joy shall be infused into the faithful operatour, whos good fortune it wil be truely to observe and anothomize wth a skilfull & expert hand, that most charitable nurse amongst the vegetables in wch this worlds great Queen & multiplicative influence appeareth so plentifully, as in her vegetable temple; for the very same shall (as in a looking glass) behould and perceave all the secret mysteries of Nature, and her ministers . . . [69]

In his experiment on wheat, Fludd showed how each stage of the operation corresponded with the stages of the macrocosmical creation, and also showed how all the elements are contained within this most noble representative of the vegetable kingdom. While the English text of this experiment, written about 1620, was never published, Fludd translated the experiment into Latin and included it as the first part of the *Anatomiae Amphitheatrum*, published in 1623.[70]

In 1631 Fludd was at work on another book to explain his philosophy, which was posthumously published in both English and Latin, the *Mosaicall Philosophy*.[71] In this book he alludes to some of his earlier work, but he also offers some new proofs by which men can know the truth of his philosophy and the errors of the Aristotelians. The most important experimental demonstrations presented are the Calendar- (or Weather-) Glass and the properties of magnetism.

The Weather-Glass was a combination barometer-thermometer and was used to show the effects upon the natural world of the dichotomies light–dark or heat–cold. These opposites were central to the dynamics of Fludd's universe, and he felt that the Weather-Glass was a convincing ocular demonstration of his theories. This instrument was made of "a round or ovall glasse, with a long and narrow neck whose orifice, or mouth and nose, ought to bee proportionable unto the rest of the neck . . . "[72] On the neck Fludd placed a scale starting with figure one in the middle and then numbering up and down in sequence. The bulb was heated over a flame to rarify the air inside, then the open end of the neck was placed in a vessel of water. As the bulb cooled, the water would rise into the glass; the figure one could be calibrated on the scale to correspond to the level at any termperature. From this instrument Fludd correctly concluded that it was the action of heat or lack of it which rarified or condensed the enclosed air, causing a corresponding fall or rise in the neck. Universalizing this concept, he states further:

> For by speculation we shall find, that there is nothing in the whole Empire of Nature, which can be rarified and made subtle, except it be by the action of light or fire, whether it be visible or invisible; and the essentiall effect of that action is light. And on the contrary part, nothing can be condensed or inspissated, where darknesse hath not dominion; forasmuch as darknesse is the essentiall root of cold, which is the immoderate actor in condensation.[73]

Observations obtained by the use of his re-invention[74] were both thermal and barometric. He noted that a sudden fall in the water level meant rain was due; if a South or East wind blew after a North or Westerly, the water would fall, and vice versa. The higher the water climbed on the scale, the more cold "has dominion in the air":

> If it mount unto the 3. of the same Hemisphere, it doth foretell a slight frost: but if it ascend unto 4. or 5. it pretendeth a hard and solid frost: if it come unto 6. and 7. it argueth great ice; but if it mount yet higher, it sheweth that a hard Ice is likely to surprize and cover the whole river of Thames.[75]

The opposite end of the scale shows the corresponding degrees of heat, "which importeth the Summer or hot Hemisphere." Fludd graphically shows the universality of his instrument in the diagram shown in Plate 10. Here the scale is made according to winter and summer poles with an "Aequinoctiall" line exactly dividing the two. This miniature cosmos is exactly equal to the larger one:

> And we tearm the place of the Aequinoctiall, the Sphear of equality, because when as the Sun is in *Aries* or *Libra*, which are the vernall and autumnall intersection of the Aequinoctiall, the daies and nights are equall; so also, the temper of each Hemisphear in heat and cold, is naturally observed to tend unto a mediocrity or equality. Even such also will the temper of the *micro-cosmicall* aire, or catholick spirituall element, be unto the earth, when the water in the Glasse is drawn up half way.[76]

Armed with the evidence provided by the Weather-Glass, Fludd can turn to an attack on the science of the Aristotelians. His prime argument is that the universe is interconnected in a unified, harmonic way; the unifying catholic bond is the spirit of God, which works "all in all," as discussed above. Fludd's first proofs are always Scriptural evidence, and on that basis alone he admonishes

> . . . that if Christian Schollars would betow that seven yars, wich they employ in their Aristotelian study, in the true, essentiall, and sacred Philosophy, they would not so erre after the manner of the Gentiles, but embrace without any rebellion or contradiction, the precepts of the true wisdom . . . [77]

The English doctor well realizes by now that Christian teachings do not seem to be able to pull the Aristotelians away from "the erroneous doctrine of their seducing Master." He knows that the followers of the "worldly wisdom" will fuss and fume about what he is saying, so now Fludd rolls out the heavy artillery, his Weather-Glass:

> But I will come unto my nearer proofs, whereby I will most evidently shew, that the doctrine of *Aristotle* is a manifest enemy, and opposite or contradictory unto the truth; which being so, it is by the Apostle *James* condemned, for a branch of that wisdom or philosophy, *which is terrene, animal, and diabolicall.*[78]

Before launching into his examination of Aristotle's precepts concerning the action of heat and cold, Fludd summarizes his own position:

It appeareth, and shall be hereafter proved out of the Book of verity, that the vertue whereby God doth manifestly operate in this world, is expressed either by attraction, from the circumference unto the center; or expulsion, from the center unto the circumference; namely, Contraction, or Dilation. For after this manner is produced Condensation and Rarifaction, whereby the heavens, and the earth, and elements, with compound creatures, as well Meteorologicall, or unperfectly mixed, and such as are compleat in their composition, were created and made.

And again, by it he operateth in this world, either sympathetically, that is, by a concupiscible attraction, or antipathetically, that is, by an odible expulsion.[79]

As opposed to this view, Fludd states the Aristotelian theory: the belief that the heat of the sun and stars is attractive, rather than expulsive:

The Peripateticks being perswaded thereunto by their Master Aristotle, do accord in this, namely, That the Winds, the Thunder, the Comets, the Clouds, and other such like Meteors, are made and caused by the attractive heat of the Sun, and other Stars, which draw up vapours and exhalation out of the water and earth, and elevateth them into the regions of the aire.[80]

But, Fludd argues, the testimony of the Weather-Glass shows just the opposite to be true; the heat of the sun is expulsive and rarifying, while the lack of it is attractive and condensing. In the Renaissance doctor's mind, there could be no clearer proof of the truth of his doctrine and the falsity of the Aristotelian.

The other major experimental evidence Fludd offered in the *Mosaicall Philosophy* was to "confirm the loving Micro-cosmicall Attraction, or sympatheticall Coition, and anti-patheticall Expulsion or hatred, by the magnetick and expulsive property of the Macrocosmicall Load-stone."[81] In the same way that the Weather-Glass could be considered a miniature universe, the attractive and repulsive activating principles could be demonstrated by the mineral lodestone. Not only does the terrestrial lodestone demonstrate the principles on the macrocosmic scale, it also shows how these properties work in the animal kingdom, and can thus be used for medical cures:

... because in the mineral kingdom, there is found nothing in al the world, so neare in virtue unto the action and life of the animal, as in the Magnet or Loadstone; ... For this reason every like particle in the animal or vegetable kingdom, that worketh after the same manner, are justly tearmed Magneticall. Hereupon the well experimented Doctor, *Paracelsus*, when he writeth of the mysticall Mummies, as well corporal as spirituall, and of the attractive means or manner to extract them, as well out of the living as dead bodie; He, for the better instruction of his Schollers, and such as he termeth *filios Artis, the children of Art*, expresseth examples, drawn from the Load-stone and the Iron.[82]

Besides Paracelsus, Fludd's main source of information about the lodestone came from William Gilbert (1540–1603). Gilbert's well-known book *De Magnete* appeared in London in 1600, and in his will he gave his entire library, globes, instruments and cabinet of minerals to the College of Physicians, where Fludd could easily have had access to them.[83] In his treatment of the mineral magnet, Fludd makes frequent reference to Gilbert, whose work he respected highly, calling him "my renowned Fellow or Collegue, D. *William Gilbert*, for his Magnetick skill, and deep search as well contemplative as experimentall ... "[84] Fludd could say this about his colleague because he did not find Gilbert's work in any way contrary to his own theories. There was another member of the College of Physicians whose work Fludd cited in his treatment of the magnet, Mark Ridley (d. 1624). Ridley published in London two studies on the lodestone, one in 1613 and one in 1617

Thus could man, the Microcosm, learn of his origins and the system by which he lives. Not only is there solid foundation for this knowledge in Holy Writ, but also it could be proved by terrestrial experimentation. This is so because everything in the universe has the same compositon in essence and operates throughout by the same divine principle of emanation and withdrawal; therefore conclusions taken from experiments with wheat, the Weather-Glass and the lodestone could be universally applied. Even with all this proof, Fludd knew it would be very difficult to persuade many away from following Aristotle:

And therefore it will be no marvell, though I shall find this mine admonition rejected, and repined at by many, though perchance more aceptable unto such as are vertuously inclined unto the

truth, and are apt, yea, and sufficient in their purer
discretions, to distinguish and separate the errours of *Aristotle*,
from the infallible verity of sacred Writ, and to carry their
judgments so justly and sincerely, that the All-hallowed honour
of the one do not suffer any detriment or indignity, by the
paganish and unsanctified axioms or assertions of the other.[85]

IV. APPLICATIONS

Let us now turn to some particulars of the workings of Fludd's
philosophical-scientific-religious edifice, which will lead to his
medical theories. There is another dichotomy used by the
Renaissance philosopher (also taken from the ancients), which
forms the other key part of his cosmological schema: that of
moist–dry. Combined with the dichotomy light–dark (variously
described as hot–cold, spiritual–material, or angelic–diabolic), the
four-fold nature of the world is formed, including its humoral
equivalent in man. Here below in the elementary world, the
regions are a manifestation of the four stages of the primary
macrocosmic element, water: fire is a hot and dry water, air is a
hot and moist water, physical water is a cold and moist water, and
earth is a cold and dry water. Each of these four can be
transformed into another by the appropriate application or
withdrawal of heat. Not only is this evident in the greater world,
according to Fludd, but also it can be demonstrated in the
chemical laboratory, as in the experiments previously discussed.

The extension of the light–dark principle of creation (appearing
in the form of heat or lack of it) acting on the primeval universal
element, the waters, formed the basis not only for Fludd's
cosmology, but also for his astrology, cosmography, meteorology
and medical pathology. As previously discussed, the action of
heat–cold on the lower regions is variously described by Fludd as
the out-and-in breathing of God, the Nolunty or Volunty of God's
will, or the negative or affirmative divine will; all of them manifest
a condensation or rarefaction of the universal primary watery
element, having a direct effect on man and his environment.[86]

Another consequence of the nature of creation is that there are
two equal but opposite hierarchies of ministers or servants. The
one is angelic, ordained to carry out the divine will; the other is
diabolic, meant to carry out the volition of the forces of darkness.

During the times of the outpouring of divine light or warmth, it is transmitted below by the angelic ministers; when there is a withdrawal, the angels become passive and inactive, while the opposite spirits become active. Here below, heat tends to unfreeze, loosen and rarify, while cold tends to congeal and thicken. Too much heat, of course, can also have adverse effects.[87]

Considering the four-fold nature of things, Fludd assigns angels to govern the four cardinal winds. The governors are supervised by the archangel Michael. The entire structure is set up this way:

The Eternall spirit of wisdome, who is the initiall principle of all things, and in whom and by whom, (as the Apostle teacheth us) *the Angels, Thrones, potestates and dominations were Created*, doth operate by his Angelicall Organs of a contrary fortitude, in the Catholick Element of the lower waters; both the effect of Condensation, and that of Rarefaction . . .

. . . this one spirit worketh in, and by spirituall and Angelicall Organs, in the execution which is effected by the property of the 4. winds . . . By which it is evident that these Angelicall Presidents over the 4. winds were the Ministers and Organs by and in the which the spirits or blasts of the winds were emitted or retained according unto the will of that eternall spirit, which guideth them when and where he list[88] (Plate 11).

By daily observation, Fludd says, we can see that the essential virtue of the northern spirit is cold and thereby contractive, or attractive from the circumference to the center, thus a cause of congealing and condensation. Conversely, we see that the southern and easterly winds cause rarefaction and "subtiliation" through the action of their heat. These hot winds undo by rarefaction that which the cold northern winds effected by congelation.

Even before they came to govern the winds, the angelic servants of the divine will inhabited the stars and planets, which influences also have a direct effect on our earthly condition (then come the winds affecting meteorological phenomena). The entire operation is summarized this way:

Whereby it is evident that the eternall Breath is that which animateth the Angels; the Angels give life and vigour first, unto the stars, and then unto the winds; the winds first informe the elements, or rather alter the catholick sublunary element into

divers natures, which are tearmed Elements; and then by the
mixtion of divers windy forms in that one element, they do
produce meteorologicall compositions, of divers natures,
according unto the diversity of the windy forms which alter it.[89]

Fludd wants to stress the unifying presence of God throughout the
entire range of the created universe, continually employing the
phrase that He is "all in all," without whose vivifying presence
acting through the aerial spirits we could not live. It is through
their action that meteorological phenomena take place:

For as much therefore as the aire is a part of the celestiall
consistence, it followeth that it was made by the Word, and that
it doth as it were swim in the Word. Forasmuch as it
comprehendeth all things . . . and it is moved and guided by the
Word; yea verily, and in the aire (being it is the universall
Treasury of God) there are many peculiar cabinets, out of which,
by his Word . . . he doth produce divers kinds of
Meteors . . . which are committed unto the government and
presidentship of divers Angells or Spirits . . . These spirits . . . do
exercise their office or Ministry in the aire, and are by Gods
Ordination conversant about the directions of Tempests,
Clowds, Rain, Snow, Hail, Frost, Lightning, Thunder, Comets,
Chasmus, Floods, or Inundations, Heat, Cold, Moysture,
Drowth, and all other Accidents which do appear in the aire.[90]

In fact, using as authority a quotation from Psalm 104, Fludd says
that it is evident that nothing at all is effected in the universe, in
heaven or on earth, unless it is performed by one of the "organicall
spirits."[91] Furthermore, following Reuchlin, all bodies, "celestiall
or terrene", have spiritual directors, rulers of their virtues and
presidents of their operations; they rule the reasonable (man) as
well as the unreasonable (the stars, animals, minerals, vegetables).
Turning to the Cabalist "Archàngelus," we find that there are
many degrees in the offices of Angels:

Some . . . do stand before the divine tribunall of God, still
praising him; some administer unto him, and unto us also; some
have the custody of the watches of the night . . . Some have the
government of the four quarters of the year, and these are
Presidents over the four stations of the Sun; others are ordained
rulers over the seven Planets; and some do dispose of the

influences and vertues of the fixed Stars, and twelve Signes, of which St *John* [Fludd's patron saint] doth seem to make mention, saying, *That in the twelve gates*, that is, the twelve Signes (as *Plato* saith) are the ports of heaven, are the twelve Angels. Some are tutelar Angels, and are ordained for the creatures safeguard; some have custody of beasts, others of plants, others of pretious stones and minerals.[92]

The differences in astrological influences are explained by Fludd in this manner: according to the Cabala, divine emanations proceed through the varied, sometimes contradictory channels or ports, thus manifesting themselves below in ways either sympathetical or antipathetical:

> For if once the ten names of God, which produce ten divers emanations, of different conditons, which are sent by the ten foresaid numerations, or sephiroticall ports or channells, do breed contrary effects, both in heaven above, and in earth beneath, it followeth by the foresaid testimony of the Prophet *David*, that there must be so many angelicall vehicles, to conduct them into the lower world, as there are Cabalisticall ports and channells; and consequently as many diversities of the divine properties, proceeding from the variety of his will, as also varieties of vehicles to conduct them.[93]

The first emanation is the Platonic World Soul, or the *Primum Mobile*; it is passed to the Seraphim; the second passes through the Cherubim and is received by the Archangel Zophiel, who directs them to the fixed stars of the Zodiac; the third emanation proceeds through the Thrones, then the Archangel Zabkiel, then down to procreate Saturn, which is replenished with spirits or Saturnine Intelligences; the fourth passes through the Dominates to produce Jupiter, the fifth Mars, sixth the Sun, seventh Venus, eighth Mercury, ninth Moon, and the last descends to the order of the blessed souls to be conducted directly to the elementary world.[94]

Thus Fludd concludes that

> . . . First, . . . God doth onely operate essentially all in all, in and over all: next, that according the variety of his Volunty, he worketh diversly in this world, . . . [and] things were created, the one either Sympathizing spiritually in affection with another, or Antipathysing among themselves, by reason of beams of a

contrary disposition, according unto the concordant or opposit nature of the Angelicall irradiation . . . [95]

Each and every creature generated here below "had his radicall information by these Emanations,"[96] that is, was distinctively made according to the particular irradiations at its hour of birth. Not only may this produce a human animal or plant offspring unlike its parents,

> . . . it may happen that the radiation of a plant, or flower, or beast, or such like thing, may for ever Antipathise with (or be disagreeable to) this, [while the same creature may] aptly agree and Sympathise with another.[97]

This point is quite important for the practice of medicine, in Fludd's view, because whereas it is apparent to everyone when there is discord between opposite elements, i.e., water–fire and earth–air, and dissonant qualities (cold–heat and moisture–drouth),

> . . . there is yet a more latent, and internall cause of Sympathy, and Antipathy in things, which by some is ascribed unto the occult natures of the Starrs: but in verity it proceedeth from those Angelicall influences, which do invisibly and after a most occult manner, stream out of creatures that are born under a discordant Emanation . . . [98]

Consequently, it is absolutely necessary to cast the chart of a patient to discover any hidden antipathy in order to intelligently prescribe a remedy. The Renaissance physician wants to emphasize, however, that this is not the same as vulgar astrology; the common "Astrologian" mistakes

> . . . the visible organ for the invisible agent, the externall creature for the internall angelicall vertue, the stary influence for the hidden super celestiall emanation, which is poured into the spirits or intelligences, which inhabit and illuminate the stars, and send it down again from them into the elementary world, to animate the winds, and by them the catholick element, after a fourfold manner, and by the element so altered, to inform the meteorologick bodies diversly, and by them the severall compounded creatures, both in the sea and land.[99]

It is the height of ignorance, in Fludd's opinion, to attempt the practice of medicine without a knowledge of the basic inner structure of things and its operating principle, the outpouring or indrawing of the divine spirit. Even though the "Ethnick" philosophers and physicians have discovered certain cures "by practicall effects, or sensuall observations, and demonstrations *a posteriori*," they can discern no other reason for the working of these cures but that they are of a "hidden property."[100] As a consequence, the only thing one can learn from such doctors is "*Ignotum per ignotius*, a thing unknown by a more unknown."[101] Aristotelians are ignorant of the core of the polarity found in the phenomenal world, so that the "true mystery of plentitude and vacuity, was utterly unknown unto the sect of the Peripateticks, because they were altogether ignorant of the true wisdom [i.e., Christian Revelation]."[102] How does this system work for Fludd's medical pathology and practice? This can be summarized as follows: creatures on earth are affected by the ever-flowing divine beams of the occult forces, both good and evil, coming from the stars and planets; these forces in turn control the blowing of the four winds, which are also both good and evil. The angelic forces of both kinds work in paired contraries as governors of the four winds. The Northern spirit is cold, therefore the North wind congeals, thickens and contracts, indicating a lessening of the divine plenitude and permitting an increase in the activity of forces of darkness; this cold spirit is driven away by the warm winds of the South, which dilate, dissipate and rarify; the Westerly wind is cold and moist, and its effects are destroyed by the hot Easterly wind.[103]

In strict adherence to the microcosm-macrocosm correspondence, the rule is: as above, so below. Thus just as there are good and evil spirits above, on earth "the whole air is replenished as well with spirits of darknesse, as with spirits of light. And therefore there is a continuall conflict made here below, betwixt those spirits of opposite conditions."[104]

The determining factor for the well-being of earthly creatures depends on which astrological influences have the greater strength; they cause a good or evil wind to blow, which in turn affect the aerial spirits:

By reason of these spirits of a contrary fortitude in the aire, sometimes good and propitious events befall the creatures of the lower world, namely when the good spirits rayne, and wholsome winds do blow, which happen, when the benign starrs and Planets have dominion in heaven, and again sometimes bad accidents . . . befall the creatures of the Elementary region, by reason of severe emissions of beams from the winds, which animate those evill spirits, that in infinite multitudes do hover, though invisible, in the aire . . . [105] (Plate 12)

Since man is constantly taking in air with all of its infused spirits through breathing (they can also penetrate the body through the pores), the agents of good or evil are likewise ingested. A normal, healthy human being enjoys a perfect harmonial balance (as in the heavens) of the four humors throughout the body, but sometimes a part of the body is unable to resist the onslaught of evil forces:

. . . if ther chance but a weakness or dibility of . . . thos heavenly and sympatheticall bonds, by which harmony and concord should be preserved . . . in one or more portions of the body, it then breedeth a malady or sickness by so much more dangerous by how much theyr harmonical bands are attenuated or violated[106] (Plate 13).

The remedy is designed to restore "that universal tye and knot in Nature whereby the foure disagreeing elements are peacably united in ye outward mass"; it is imperative to halt the disintegration of the microcosmic harmony before disaster ensues:

For as we see that one unruly horse in a stable beinge brooke loose never seaseth to jarr and fight with the rest of his companions til he remaine sole victor in the place, even so one element feeling the want of his quintessentiall bridle or ligament, breaketh forth from his ranke, and never leaveth to persecute his contrary in the same composition till all the aeconomicall habit of mans palace be destroyed and the life's spirit set at liberty.[107]

Harmony and balance are restored "by nourishing and refreshinge of those faintinge or debilitated tyes of Nature by theyr like . . . "[108] Here Fludd, while using the ancient humoral system, clearly breaks with the Galenists in using the treatment of opposites, and advocates a treatment of "like unto like." By this he means that in treating a patient, it must be determined which evil influences are at work, so that their equal but benevolent spirits can be administered as a counteraction. For this reason it is absolutely necessary to draw up an astrological chart, as Hippocrates urged, to see which planetary agents are at work; their influence is countered by herbs or minerals, also chosen astrologically, which provide an infusion of angelic forces of the same nature as those causing the malady. Examination of the urine, feeling the pulse and listening to the patient's complaints only serve to refine the testimony of the astrological chart (Plates 14–16).

V. SUMMARY

Although this is a necessarily brief summary of a long, intricate and detailed system, which Fludd worked on for years and spread over many published and unpublished pages, the above should provide some of the major aspects of his philosophical framework. Throughout his works Fludd remained quite consistent, only refining minor details from time to time. It is also apparent that this late Renaissance philosophy based on Scripture, some ancients, notably Plato, and the Renaissance revival of Neoplatonic Hellenistic texts, was particularly vulnerable when the thinkers of the seventeenth century began to separate science from religion and ancient authority, physics from metaphysics, spirit from matter. Without a belief in an all-encompassing angelic structure of the universe and a Neoplatonic emanative harmony of that angelic structure, Fludd's whole system collapses rather easily. The beauty of such a system lies in its being systematically all-encompassing, consistent throughout, in concord with many venerated ancient authors, both religious and philosophical, and in its making man's place in the whole scheme completely clear. It was just as universal a *Weltanschauung* as the Aristotelian one it was meant to replace, although both had a number of points in common. Fludd's system was also the epitome of the Renaissance Ideal: to revive ancient texts and absorb the wisdom therein to purify every aspect of human society in order to relieve the present condition of ignorance and suffering. It is within the context of the Renaissance Hermetic tradition that he succeeded so well, in his view, in achieving that ideal. That many others did not share his view in his day and beyond because of their own notions of how such a program should be carried out, relying first on the ancients, and then passing to the "moderns," is quite evident. But no other Renaissance thinker produced works of such monumental scope, with illustrations of such quality and quantity; and although it has been pointed out in this study and others that there were a number of those who disagreed with Fludd in varying degrees, there were also a considerable number of scholars, physicians, noblemen, clergy of various denominations and levels, artists, poets and writers of the age who shared much of Fludd's speculations and those of his antecedents. The pioneering work of Dame Frances Yates and others has served to show some of the interrelationships and influences on the Renaissance Hermetic-Cabalist tradition, and suggest others. Therefore, if we are to have a full understanding of, and appreciation for, a time which produced the complex poetry of Shakespeare and the King James Version, among other surviving achievements of the era near the end of the ancient world and the beginning of the modern, Fludd and his works must be considered important for accomplishing that task. Because his books and papers covered the full range of human thought and endeavor, they may be fruitfully mined for insights for many years, and his illustrations will no doubt continue to intrigue and delight new generations of viewers.

Chapter IX

THE ROSICRUCIAN CONNECTION

And thence proceedeth that fair concord, that, as in every several kernal is contained a whole good tree or fruit, so likewise is included in the little body of man the whole great world, whose religion, policy, health, members, nature, language, words and works, are agreeing, sympathizing, and in equal tune and melody with God, heaven, and earth. And that which is disagreeing with them is error, falsehood, and of the Devil, who alone is the first, middle and last cause of strife, blindness, and darkness in the world.

Fama Fraternitatis, 1614

During the English Civil War, Thomas Fuller wrote, in his widely-read *The Worthies of England*, that Robert Fludd "was of the Order of the *Rosa-Crucians*," and in 1691, Anthony Wood's *Athenae Oxonienses* states that "He was a zealous brother of the order of Rosicrucians . . . "[1] In more recent times, J. B. Craven titled his 1902 biography *Doctor Robert Fludd: The English Rosicrucian*, in the third chapter of which he had Michael Maier, the German physician and the other main Rosicrucian apologist besides Fludd, initiating the London doctor into the Fraternity during his visit to England.[2] In 1924, A. E. Waite, in his *Brotherhood of the Rosy Cross*, say that "I am quite certain that he [Robert Fludd] did not found the order of the Rosy Cross, but he may have belonged to something at work under that name, perhaps in 1617, perhaps later."[3] Dame Frances Yates, in *The Rosicrucian Enlightenment* (1972), terms Fludd a "Rosicrucian" philosopher but in *The Occult Philosophy in the Elizabethan Age* (1979), the quotation marks are dropped: "Milton inherited the Neoplatonic Christian Cabala of the Elizabethan age as expressed by its early seventeenth-century successor, Robert Fludd, the Rosicrucian."[4]

From the above statements, it is clear that any study of Fludd must necessarily attempt to deal with two questions: who the Rosicrucians were (or were not), and whether he was a "brother" of such an "order," or otherwise what his connection may have been.

On the first question, there have been mountains of pages written over more than three centuries, most of which, unfortunately, has done little to clear up the matter. Much of this writing has ranged from unsubstantiated claims to totally credulous acceptance of the literal truth of the Rosicrucian myth put forward in the original manifestos. Some recent scholarship has helped shed some small measure of light on the first question[5] (discussed further below): on the second, we can be a bit more certain, since we have the testimony of Fludd's own words in the matter. I propose that there was no Rosicrucian Brotherhood of any significance as described in the manifestos of 1614 and 1615, and that it can be shown with some certainty that Fludd was not a member of such a brotherhood.

While he was a student at Oxford in the 1590s, Robert Fludd already began to synthesize some of the subjects of his studies which would appear some twenty years later in the *History of the Macrocosm*. It was there he composed his treatise on music, which was central to his entire Pythagorean-Neoplatonic cosmology; and he was already well-versed in astrology. About 1601–2, he wrote the treatises on arithmetic, geometry, perspective and "the secret military arts" for the private tutoring of Charles of Lorraine, fourth Duke of Guise, and his brother François, a Knight of Malta. The discourses on the art of motion and on astrology were composed about the same period. All these tracts were pulled together in a completed *History of the Macrocosm* about 1610–11, which was then read by others in manuscript. After that, the industrious physician probably went to work on the sections of the *History of the Microcosm* and other tracts of interest, which he dictated to his secretary. Thus by 1614, Robert Fludd was a forty-year-old, well-established London physician who had completed in manuscript a wide variety of tracts, which were organized according to twenty-some years of thinking and writing about his Christian, Cabalistic, Alchemical, Neoplatonically harmonious universe.

That same year, the Rosicrucian phenomenon began, initiating over three centuries of unanswered questions and controversy which has continued to this day. It started with the appearance of the two supposed original manifestos of the Brothers of the Rosy Cross, the *Fama Fraternitatis* of 1614 and the *Confessio Fraternitatis R.C.* of 1615. Both came from the press of Wilhelm Wessel in Cassel, Germany.

The first edition of the *Fama Fraternitatis* contained, in the first section after the Epistle to the Reader, a tract entitled the "Universal and General Reformation of the Whole Wide World." It was known as early as 1618 that the "Universal Reformation" was a German translation of an Italian work, later identified as a section from Trajano Boccalini's *Ragguagli di Parnaso* of 1612.[6] This particular "Report from Parnassus," number 77 of the first century (there were two altogether), is a delightful, hard-hitting satire about the inability of those in authority to be able to effect a general reformation of the world by having to rely on the learned establishment. The Roman Emperor Justinian presents Apollo, at court on Parnassus, with a new law against suicide for his approval. It seems that the world is in such a wretched condition that many people were choosing to leave it early. Apollo resolves to call a council of the wise to deal with the problem, but

> . . . amongst so many philosophers, and the almost infinite number of vertuosi, he could not find so much as one who was endowed with half the requisite qualifications to reform his fellow-creatures, his Majesty knowing well that men are better improved by the exemplary life of their reformers than by the best rules that can be given.[7]

The ruler decides to proceed anyway, charging the Universal Reformation to an assembly of the Seven Wise Men of Greece, Marcus and Annaeus Seneca, and the contemporary Italian philosopher Jacopo Mazzoni. After debating and rejecting a number of far-fetched schemes, the assembly determined that they should call the Age itself before them for a complete examination and diagnosis. After scraping away the infested appearances covering the patient's body, no healthy flesh could be found, so he was dismissed as a hopeless case. This being so, the learned gathering knew that it was still expected to come forth with a reformation of some sort; so they gave up considering public affairs and closeted themselves to come up with something which would save their reputations. Thus they produced a Manifesto of a General Reformation which "fixed the price of sprats, cabbages and pumpkins," and, as an afterhtought, included peas and blackberries; then,

> the palace gates were thrown open, and the General Reformation

was read, in the place appointed for such purposes, to the people assembled in great numbers in the market-place, and was so generally applauded by everyone that all Parnassus rang with shouts of joy, for the rabble are satisfied with trifles, while men of judgment know that *vitia erunt donec homines* [Tacitus, Hist., iv.] – as long as there be men there will be vices – that men live on earth not indeed well, but as little ill as they may, and that the height of human wisdom lies in the discretion to be content with leaving the world as they found it.[8]

At first glance, this satirical piece may seem a curious introduction to the first Rosicrucian manifesto, which appears to have a general reformation as its aim. The "Universal Reformation" has variously been dismissed as unrelated to the manifestos, taken as proof of the Rosicrucians as a hoax, or left as a source of puzzlement. As discussed below, it is my opinion that the placing of the "Universal Reformation" before the *Fama* was both deliberate and purposeful and that it is entirely consistent with the theme of the *Fama*.

Following the "Universal Reformation" in the 1614 edition was the *Fama Fraternitatis of the Meritorious Order of the Rosy Cross, Addressed to all the Learned and the Leaders* [Haupter] *of Europe*. This tract weaves a mythological tale about an initiate into the divine mysteries, Brother R. C., his founding of a Fraternity of the Rosy Cross, and the present revelation of the order after the rediscovery of his wondrous tomb.[9]

The purpose of the manifesto is clearly set out at the beginning: to renew and perfect the arts of man so that he "might thereby understand his own nobleness and worth, and why he is called Microcosmus," and how far his knowledge of Nature may extend. This has not been accomplished heretofore because of the "pride and covetousness" of the learned, who cannot bring themselves to agree for fear of losing their prestige and standing. Instead of gathering the God-given riches provided in the Book of Nature, which would raise all arts to perfection, they continue to adhere to "Popery, Aristotle, and Galen, yea and that which hath but a mere show of learning, more than the clear and manifested light and truth."[10] Indeed, if Aristotle and Galen were now living, they would abandon their erroneous doctrines. However, even though the truth concerning theology, physics and mathematics is now known, the prince of darkness continues to hinder the truth through those doing his will, which include "contentious, wavering people."[11]

It was to institute this general reformation that Brother R. C. founded the Fraternity of the Rosy Cross. He had learned the divine secrets of nature from the wise men of Arabia and North Africa, but when he offered to share his secrets with the scholars of Europe, they laughingly dismissed him, "and being a new thing unto them, they feared that their great name should be lessened, if they should now again begin to learn and acknowledge their many years errors, to which they were accustomed, and wherewith they had gained them enough."[12] As a result, he retired to a solitary life of contemplation in Germany. It was noted that Paracelsus, who had carefully read the "book M" (which Brother R. C. had translated from Arabic) even though he was not a member of the Fraternity, had met a similar fate in being obstructed by "the multitude of the learned and wise-seeming men" in attempting to share his understanding of nature with others.[13]

Five years after withdrawing to Germany, Brother R. C. decided to actively take up the cause of a general reformation once again. He thus started the Fraternity of the Rosy Cross, first by initiating three members, then seven, who each went into a different country in order to be the bearers of light and truth. They would be able to impart their knowledge to the worthy learned in secret, and observed if there were erroneous doctrines which should be reported.[14] In addition, they would be available for wise counsel to those who needed it, in a similar fashion to the ancient oracles, and serve the even more important role of tutor to future rulers. For the present rulers, they would produce "gold, silver and precious stones . . . for their necessary uses and lawful purposes" (probably metaphors for wisdom).[15]

The brothers all agreed to six rules: 1) Their only profession was to heal the sick, *gratis*; 2) To wear the customary clothing of their host country; 3) To meet annually on the day C. in the house *S. Spiritus*; 4) To seek a worthy successor; 5) Their seal and mark would be the word C. R.; 6) The Fraternity was to be kept secret for one hundred years. And thus they lived and worked, each initiating a successor before his death, until the year 1604, when the unknown burial chamber of Brother R. C. was found.

The discovery of this marvelous tomb, which was replete with allegorical shapes, dimensions and objects representing the harmony and wisdom of the cosmos, including three books of Paracelsus, marked the end of keeping the order secret. It was now time to push forward with a general reformation, "both of divine and human things," even though some progress had been made in the previous hundred years. The brothers therewith declared that they were reformed Christians (probably Lutheran), and recognized the Roman Empire and Fourth Kingdom as their polity. As for their philosophy, it is not a new one, "but as Adam after his fall hath received it, and Moses and Solomon used it . . . And wherein Plato, Aristotle, Pythagoras and others did hit the mark, and wherein Enoch, Abraham, Moses, Solomon did excell, but especially wherewith that wonderful book the Bible agreeth."[16]

They also soundly condemned "ungodly and accursed gold-making", which had become so pervasive, and declared that true philosophers esteem that little, since they have much greater things to pursue, though they have knowledge of the art. Indeed, there were many chemical books in circulation which were *in contumeliam gloriae Dei*, and the pure of heart should beware of those listed in a catalogue furnished by the fraternity.

Accordingly, those of the learned community reading the *Fama* and *Confessio* who wished to join in this great reformation should express their wishes, singly or jointly, in print. All those earnestly seeking their attention would assuredly be considered; the upright would thereby benefit both materially and spiritually; but the greedy and false-hearted would only bring upon themselves "utter ruin and destruction," while being completely unable to bring harm to the fraternity.

In the 1614 edition of this work, the *Fama* is followed by a laudatory reply to the Brothers by an Adam Haselmayer, which was reportedly first printed two years earlier. Haselmayer said that he had seen a manuscript of the *Fama* in 1610 in the Tyrol, and he includes some pointed comments against the Jesuits. The full title of the volume and the preface states that for his efforts he was seized by the Jesuits, clapped in irons, and put on a galley.

The appearance of the *Fama* caused a considerable uproar in the European community, and the 1615 edition of the *Confessio* continued to fan the flames. This volume contains two tracts, the first being the *Consideratio Brevis* ("A Brief Consideration of a more Secret Philosophy") by "Philip à Gabella," which was obviously based on John Dee's *Monas Hieroglyphica* (1564). It reiterates and expands on the glyph, invented by Dee, which represents all the symbols and processes of the universe.[17]

The second tract was the other Rosicrucian manifesto, the *Confessio Fraternitatis R. C., Ad Eruditos Europae*. The *Fama* was first printed in German, while the *Confessio* was in Latin, presumably to reach a more learned readership, as indicated in the title. It starts with a preface to the reader in which it is declared that within are given thirty-seven reasons concerning the purposes and intent of the order (which are not precisely enumerated), and the reader is thereby asked to consider whether they are sufficient to induce him to believe in their cause. Then it is declared that

> . . . we do securely call the Pope Antichrist, which was formerly a capital offence in every place, so we know certainly that what we here keep secret we shall in the future thunder forth with uplifted voice, the which, reader, with us desire with all thy heart that it may happen most speedily.
>
> Fratres R. C.[18]

The *Confessio* itself opens by saying that regardless of what one heard about the *Fama*, the message should not be hastily believed or willfully rejected. Indeed, it is Jehovah who is behind the present revelations, which previously were discovered only by long and difficult labor, because of the decayed state of the world, and "near to its end, doth hasten it again to its beginning." And in spite of the fact that the *Fama* sufficiently spelled out the nature of the Fraternity, it was deemed necessary to further explain their purposes to the learned community, so that the Brothers might more readily receive their approval. Once again, they deny being involved in any religious heresy or attempts to undermine civil authority, and restate their position:

> . . . we hereby do condemn the East and the West [meaning the Pope and Mohamet] for their blasphemies against our Lord Jesus Christ, and offer to the chief head of the Roman Empire our prayers, secrets, and great treasures of gold.[19]

The document goes on to restate their basic premises, but in more expository terms rather than by mythological tale. It gives the birthdate of Brother R. C. as 1378, and attributes to him a life-span of one hundred and six years (therefore d. 1484; the tomb was discovered 120 years later, 1604), and is much more violently anti-papal than the *Fama*, emphatically prophesying the Pope's downfall as a result of their movement. It closes with an invitation to the like-minded to join with them in partaking of their spiritual treasures.

By 1616, a torrent of activity in the literary world was keeping the presses working overtime with replies, entreaties and counter-blasts to the Rosicrucians. That same year a book appeared in Strasbourg which seemed to be another Rosicrucian publication: *The Chemical Wedding of Christian Rosenkreutz*. This long, complex mythological tale, constructed of alchemical symbolism, written by the Lutheran clergyman-scholar Johann Valentin Andreae (1586–1654), was thought then, as through the centuries to the present, to be a continuation of the Rosicrucian legend. It has also been alleged that Andreae was the author of the *Fama* and *Confessio*, and, short of that, certainly one of the original collaborators in putting forward the first manifestos. However, the exhaustive scholarship of John Warwick Montgomery has convincingly shown that Andreae, though an admirer of Paracelsus and an adherent to the Macrocosm-Microcosm view, was quite opposed to Rosicrucianism as set out in the *Fama* and *Confessio*. In fact, the *Chemical Wedding* was an attempt at realigning alchemical allegory directly with the central truths of Christianity, rather than have such knowledge transmitted secretly from the East to the West through pagan sources. Furthermore, Andreae specifically condemns the Rosicrucian fraternity and all similar occult movements of the day.[20] Thus for the purposes of this chapter, the *Chemical Wedding* need not be considered further, since it was not one of the "genuine" original manifestos, and indeed was not referred to by Fludd.

Among the avalanche of favorable apologies, replies and entreaties for consideration for membership in the illuminated order, there were also, predictably, negative responses from those who did not share a Protestant, Paracelsian-Hermetic world-view. Some of these came from the pen of Andreas Libavius (1540–1616), an independent-minded German Lutheran chemist, who had long established his reputation by authoring a monumental plain-language textbook on chemistry and chemical medicines and had engaged in the chemical debates of the day. He assailed the brotherhood in several works on a variety of grounds, which included heresy, sedition, and use of diabolical magic to carry out their aims.[21]

When Libavius' *Analysis of the Confession of the Fraternity of the Rosy Cross* (Frankfurt, 1615) came to Robert Fludd's attention, he was moved to pen a quick rebuttal in order to have it listed at the spring, 1616 Frankfurt book fair. This short piece (23 pages, octavo), the *Apologia Compendiaria*, or *A Brief Apology for the Fraternity of the Rosy Cross, Tainted by the Stain of Suspicion and Infamy, Which is Washed Away and Cleansed by a Flood of Truth*,[22] takes issue with the German iatrochemist on all counts. Fludd says that all things professing to be of the spirit and prophetic should be tested, and those which pass should be upheld. The Society of the Rosy Cross demonstrated an admirable knowledge of divine and natural secrets in their manifestos, so why not investigate them, and, if discovered, approach them, since they have stated in their writings that they will offer their gifts of wisdom freely. As for Libavius' *Analysis*, Fludd finds more resentment and maliciousness in it than diligent inquiry. Since the mother of sedition is ambition, and the Brothers have said that they desire the divine life and hold worldly things in contempt, they have sufficiently cleared themselves of this charge. And certainly their adherence to Christ and the Christian life should take away all suspicions of their engaging in diabolical arts, since their works can only be performed through the divine spirit. It is also absurd for them to have announced such things publicly if they can only carry them out by evil magic, because in every Christian country necromancers, sorcerers and conjurers are punished by pillory or fire. As for the charge that the Brothers have fallen into heresy because they wish to reform the arts along the lines of the ancient divine wisdom, it comes mainly from the constantly quarreling theologians and Papists.

Because Libavius and others of like mind seem set in their erroneous opinions, Fludd wants his short Apology to be simply an announcement of a longer work to deal with these important issues at length, which will be ready for the Frankfurt book fair the following year, which he outlines. In it he will prove, contrary to Libavius and others, that the wondrous things promised by the Fraternity are achievable without resort to diabolical magic. The expanded Apology will be divided into three parts: one, a discussion of whether Magic, Cabala and Astrology are legitimate arts or mere superstition; two, a demonstration of the major defects remaining in the arts as taught in the schools: this includes the Physical Arts (Natural Philosophy, Medicine and Alchemy), Mathematics (Arithmetic, Music, Geometry, Optics and Astrology), the Moral Arts (Ethics, Economics, Politics and Jurisprudence) and Theology; three, the arcane mysteries of nature, including occult properties and wondrous effects: for example, that it is possible to confer with another person who is hundreds of miles away by either voice or appearance, or both, and also by writing, and that this is done without the help of the Devil, indeed, by natural supercelestial means; that it is possible for someone who has written a single book to know what is in all books; that there is a great power of the true and arcane music in all of creation, which contains infinite and wondrous secrets behind the exterior of even the smallest things.

The *Apologia Compendiaria* ends with an epilogue by the author addressed to the Brothers of the Rosy Cross. Greeting them as "Brothers most dear," and in the name of Jesus Christ, whom they serve with pure and sincere hearts, Fludd recalls their promise to replace the misery from the fall of Adam with greater good fortune. He begs their forgiveness if he may have lapsed into error in their defense because of his own insufficient knowledge. He would like nothing better than to be accepted as the lowest of members in the order, so that he would be able to satisfy the inquiries of the curious in a more certain manner.

To make sure that the Fraternity understands who he is and what his beliefs are, he briefly gives these specifics: his name is Fludd, he was born of noble status, and is the lowest rank of physician in London. His wife is the desire of wisdom, and his children are the fruits of that desire. The body is a prison, and the pleasures of the world are but vanities and destructive of the mind. Fludd says that he has examined, with his eyes and his soul, almost all the countries of Europe; the raging deep seas, the high mountains and the slippery valleys; the rudeness of villages, the incivility of towns, the arrogance of cities; and the ambition, avarice, infidelity, ignorance, idleness, cunning and almost all the wretched aspects of mankind; and he has found no one who rightly knows himself in accordance with St John 1: "the life was light, and that light was in darkness, and the darkness comprehended it not." Indeed, what he has found is vanity reigning triumphant everywhere. And then he concludes with a standard Latin closing, bidding the Brothers farewell and to be mindful of him.

Just as promised in the short Apology, Fludd produced a longer one (196 pages, octavo) by the time of the 1617 book fair. The *Tractus apologeticus integritatem Societatis de Rosea Cruce defendens*[23] exactly carried out the plan presented in the small book of the previous year, and the *Apologia* (less the outline) is included in the 1617 work as a Proemium. Quoting specific passages from the *Confessio* and Libavius' *Analysis*, the apologist maintains that Magic, Cabala and Astrology are true arts if rightly done, and are to be distinguished from the black arts of a necromantic and diabolical sort. Indeed, it is imperative for anyone wishing to understand the occult sciences to gather true knowledge from the Book of Nature, since that provides the necessary divine wisdom and insight. As for the teaching of Natural Philosophy, Medicine, Alchemy, Mathematics and the Moral Arts in the universities, it is in a disgraceful condition and thoroughly in need of reform. The defects of each particular category are detailed, which center around the heavy reliance on a mere descriptive approach to nature. In drawing on the writings of Aristotle and Galen, the arts and sciences are maintained at a dangerously superficial level, while the true essence of things is ignored, when in fact the arcane secrets of nature should be sought after, studied and taught, since therein lies the divine light and therefore true wisdom. In fact, the key to knowing the interrelatedness of all things in the cosmos, great and small, comes from an occult understanding of the musical harmony throughout the universe. In the last section, he describes how the divine light is the central animating force in the created universe, and how all sorts of wondrous effects may be manifested, all of which are done through the divine principle and without diabolical aid.

In Fludd's view, therefore, his own orientation is totally in accord with the statements found in the Rosicrucian manifestos, so the Fraternity certainly should not be dismissed out of hand, but approached and investigated. The *Tractatus Apologeticus* finishes with the same epilogue addressed to the Brothers as that at the end of the earlier short Apology, with an addition which could be translated as:

Farewell, Brothers most dear, in the name of those whom you

sincerely honor. Farewell, I say, and farewell again; favor me and approach me (I implore and entreat you by your assurance and because of the ignorance of the age in the true and pure Philosophy); be mindful of me and your promises.[24]

It is apparent that Fludd believed, as so many others did, that the Fraternity was real, and that he would make an ideal candidate; he was a celibate bachelor; a physician who was probably both willing and financially able to heal the sick *gratis*, particularly in exchange for the great wisdom to be obtained from the Brothers; his own philosophy was completely in line with theirs; and he had eloquently risen to their defense in his Apologies. Certainly the two Apologies (both published in Leyden by Godfrey Basson, the son of an Englishman and lover of occult subjects) had the effect of drawing the attention of many to the author from that time on, to his benefit or detriment. Unfortunately, however, they did not draw the attention of the Brothers themselves, which was the common experience of all the aspirants.

The same year as the appearance of his second Apology, 1617, two more of Fludd's books were published, this time by the firm of Johann Theodore de Bry, the Palatinate and Frankfurt publishing and engraving house with which Fludd enjoyed a long (1617–1633) and exceptionally fruitful relationship.[25] The shortest of the two was the *Tractatus Theologo-Philosophicus*, a work which could be considered in a sense as another Rosicrucian Apology, although the Roscrucians only occupy a little over ten of 126 pages. This tract, which is dedicated to the Fraternity, deals, in its three sections, with Life, Death and Resurrection as its main topics, and repeats the familiar themes of Fludd's construction of creation, with the light–dark, divine–evil, life–death dichotomies derived from Scriptures and Hermes. In Chapters XV and XVI of the first section, on Life, he discusses the fact that the arcane mysteries of the Patriarchs, Prophets and Apostles remained among men, and that there is a terrestrial paradise as well as a celestial. He goes on to say that admirable knowledge of Moses and Elias, which was the key to the hidden wisdom, remained among certain elect of clean and pure heart. These sons of God have received the Hidden Manna, the morning star, superhuman powers, the white raiments; they are pillars in the temple of God, and their names appear in the Book of Life. To the ordinary world, these virtuous men appear

poor and unknown, but they are wealthy with the abundance from
heaven, and live in their secret houses of the Holy Spirit. They
certainly are not the conventional theologians or, for that matter,
the Pope, since the real and efficacious gifts of the spirit are those
enumerated in 1 Corinthians 12: prophecy, miracles, tongues,
interpretation of tongues, healing, etc; and the real bringers of the
occult truths must be able to demonstrate prophecy, true visions,
casting out of demons, and other marvels, as well as themselves
observing the divine principles.

At this point, Fludd turns his narrative into an address to "the
most illuminated Brothers:" he has diligently scrutinized the *Fama*
and *Confessio* with his eyes and his soul, and concluded that they
are in fact illuminated by the divine spirit, and that the Brothers
have been shown the things which were prophesied mystically in
the Scriptures as preceding the end of the world. If their actions
are in consonance with their words, which Fludd is no longer able
to doubt, then their predictions must be believed, since they
conform to sacred truth. In order to illustrate again the wonders
which can take place through the divine spirit, Fludd quotes the
story of Elias at Mount Horeb from 1 Kings 19. He also lauds the
Brothers for being able to make gold, but disdaining the earthly
kind for the heavenly, which again shows their spiritual orientation.
At the end of Chapter XVI, he issues an invitation for all of
humanity, whose minds are befogged with ignorance, to join with
him in acknowledging that the Fraternity of the Rosy Cross is no
doubt under the guidance of the sacred spirit, and that their
dwelling is at the summit of a high mountain, where those who live
may inhale the most sweet and rarefied aura of the Psyche, which
is the adornment of true wisdom.

The narrrative then climbs down from these intoxicating
mystical heights to continue the section on Life, and proceed
through the ones on Death and Resurrection. In the seventh
chapter of the last section, the gold of the impious and spurious
alchemists is compared to the heavenly kind spoken of by the
Brothers in their manifestos; and the great mystery of the
resurrection is said to have been shown by the most illuminated
Fraternity in the description in the *Fama* of the tomb of Brother
R. C. The sun in the ceiling of the tomb represents the
incorruptible wisdom of Christ, shining down on the founder's
undecayed body. After this brief mention, the little book is

completed without further reference to the Order.

Since they occupy so little space in this small treatise, it is possible that it may have been a work in progress when Fludd became involved in defending the Rosicrucians, and therefore he conveniently wrote them into his scheme. Certainly here he has pulled out all the stops and fully committed himself not only to the reality of the Order, but also to their being of the elect illuminati, whose public appearance foreshadowed the end of the world as prophesied in Scripture. Perhaps this is why the author of the book was given as "Rudolfo Otreb, Britanno," an obvious anagram for "Roberto Fludo." He had already published two other Apologies for the Rosicrucians under his own name, but they both took the approach that it was possible for the Brothers to know and do the things they claimed, and therefore they should be investigated and given a fair chance to prove themselves. In the *Tractatus Theologo-Philosophicus*, however, all the doubt and tentativeness is gone, which led some commentators to believe that the difference is due to Fludd's having been initiated or apprenticed to the order by that time. This notion is directly contradicted by Fludd's own statements in his unpublished manuscripts, as shown below.

The third and by far the most important of the books published by Fludd in 1617 was the first installment of the *History of the Macrocosm and Microcosm*. This initial volume was the first tractate on the Macrocosm, which dealt with creation and metaphysics, and was dedicated to God and James I. The second tractate on the Macrocosm, which detailed the arts and sciences of man, portrayed as the "ape of nature," was issued the following year, 1618 (Plates 18 and 19). This latter year was an important one for the mystical philosopher, and it is to it and the following year that we can look for clues about Fludd and the Rosicrucians.

As discussed earlier, in 1618 Robert Fludd was elected a Censor of the College of Physicians, and on 12 May two monopoly patent holders for steel-making complained to the Privy Council that the physician was illegally making steel in his chemical laboratory. Apparently later that same year or early the next, "some envious persons" told the English King that Fludd had dedicated a large work to him, that the work contained spurious philosophy, and that the author had defended the Rosicrucians, who were guilty of religious innovation and heresy. Fludd's subsequent meeting with the King to defend himself proved so satisfactory that from then on he enjoyed the patronage of the monarch, including the King's backing for receiving his own patent for the making of steel. At their meeting, the King apparently suggested that Fludd might wish to commit his rebuttal to paper, and so he produced the unpublished "Declaratio Brevis" to James I (see Appendix A). Written about late 1618 or early 1619, the "Declaratio" states that his *Tractatus Apologeticus* in no way deals in heresy, since it defends a brotherhood with legitimate claims, and that they themselves are clearly of the reformed Christian religion (probably Lutheran). The physician adamantly maintains his own unshakable adherence to the Anglican Church, the unstained nature of his own character, and his loyalty to the King. In fact, his great work on the Macrocosm, which makes "the most evident explications and most clear demonstrations of the secrets of nature, which have been concealed or hidden by the ancient philosophers under the guise of allegorical riddles and enigmas,"[26] was dedicated to James because, second only to God, to him alone should the honor go for the fruits of his labors. In fact, a controversy had arisen between the person who carried the work from England to Germany for publication and the printers and engravers. The courier wanted to dedicate it to his own sovereign, the Landgrave of Hesse, while the publishers wished to give that honor to the Count Palatine, but Fludd stepped in and insisted on God and James I. To lend weight to his testimony, Fludd includes passages from letters of support from Continental colleagues.

As outlined in Chapter IV, Fludd thought, as so often seems to be the case, that he had not sufficiently explained his ideas, so he followed the "Declaratio" with another unpublished manuscript, "A Philosophical Key."[27] This treatise is much longer than the previous one, and affords us some quite interesting information about the early seventeenth-century mystic particularly in relation to the Rosicrucians. This tract, which was probably written about 1618–20, was also dedicated to James, his "best Patron and Worthiest Maecenas," to whom he wishes to provide a "key to unlock and open the meaning of that Macrocosmicall and Microcosmicall Philosophy" which he had also dedicated to the monarch. The "Key" contains three sections: "A Preface to the unpartial Reader;" "A Nosce te ipsum to the malitious Detractor or the Calumniators Vision"; and "Of the Excellency of Wheate." It ends with a short Epilogue to the reader. In the last section, which occupies the bulk of the manuscript, Fludd describes his alchemical experiments with wheat, and how they provided him with the basis for his philosophy. The "Calumniators Vision" contains a mythical account of man's place in the cosmos, and was described in the previous chapter.

It is in the Preface that we are provided with a number of insights about the Fludd-Rosicrucian problem. He says that he is still nursing "the green and fresh wounds of a certain silly and poor Apology of mine," which he wrote against Libavius in order to see how well his philosophy would fare in the world, and also "to insinuate into those learned and famous theosophists and philosophers acquaintance intitled the Fraternity of the Rosy Cross."[28] What he got for his efforts were malicious attacks on his religious beliefs, philosophy and character from some who later confessed they had not even read his philosophy, and others who, though they read it, had not understood a tenth of it. He was upset about a learned and lucid gentleman who maintained in his presence that Fludd had totally accepted the assertions of the Fraternity, whereupon the maligned philosopher showed his colleague a number of passages in his *Apology* where he had continually qualified his statements with phrases such as, "if they be free from devlish superstition, if their deeds do agree with their words," etc. This evidence satisfied the gentleman, who was thereby moved to acknowledge his error.[29] Interestingly, "Rudolfo Otreb's" theological-philosophical tract of the same year as the *Apology*, which totally accepted the Brothers as the spiritual heirs of Moses, is nowhere mentioned.

Fludd then takes up the charges leveled at his Macrocosm-Microcosm histories. After complaining about the "bitter assaults, sharp censuring darts and uncharitable speeches" which they evoked, the same letters from abroad as in the "Declaratio Brevis" are paraded to show his support there. As noted earlier, some had even maintained that he must have had help from the Brothers R. C. in writing his volumes, because they held it to be impossible that one could treat of "such mystical learning . . . and to be conversant in so many sciences, except other notable heads were joined in such a tedious business." Indeed, a person who seems a "great scholar and Doctor" put this concept into the head of "an honorable personage and peer of this realm." But he assures them and the world on his faith and reputation that he has borrowed no man's assistance, and that his philosophy is a result of his own invention, both "in practice and speculation." Indeed, this is the reason he has taken the trouble to detail his experiment on wheat in English, so that everyone who reads it may better understand him, and see the falsity of his accusers.[30]

He then makes the following significant statements, which deserve to be quoted in full:

> I do boldly assure the jealous world before His sacred presence, who made it, that my Macrocosmical history as well natural as artificial was composed by me some four or five years before the renoun and fame of the Fraternity of the Rosy Cross had pierced mine ears, as by the testimonies of my worthy friends Mr. Dr. Andrews, and that most learned gentleman of the Inner Temple, Mr. Selden it will easily be justified.
>
> Acknowledging even from the central essence of mine inward thoughts, that, if the said Fraternity of the Rosy Cross be so excellently informed in the mysteries of God and Nature as Europa declares them to be, my works are as far from wonder or desert in respect of theirs (being only builded on the sandy ground of opinion and outward experience) as heaven is distant from the earth, or light from darkness.
>
> Professing also unto every one before the everliving God (who knoweth each man's secret imaginations) that to this hour I neither have unto my knowledge ever seen, known or conferred with any of the said Fraternity so admired abroad of many, neither understand I of their orders or conditions more than I have gathered out of such public books as I have read.
>
> And thus much do I utter unto my grief, openly acknowledging that I have earnestly desired their society and familiarity for the better erecting and exalting of my mind in the knowledge of mysteries as well divine as natural; but finding now by so long trial that mine apology made in their defence hath no way moved their affections for me, neither hath had that magnetical force, as to allure them unto my acquaintance and conference in all this space, I do firmly imagine that they perceive some notable unworthyness in me to participate in their high mysteries, and therefore do, as it were, now despair to attain unto that, which so many thousands besides myself have both privately and publicly desired, and in conclusion failed with me in their wishes (as by daily discourses and books it may well appear).[31]

151

Particularly since the pious doctor made these statements in the form of an oath, there can be no doubt whatever about their veracity, and on that foundation their significance rests in providing a concise summary of the problem and its outcome in the clearest terms.

Again for Fludd, in the "Philosophical Key" as in the *Apology*, there does not seem to be any doubt that the Fraternity exists, and an equally important point is that he believes their miracles to be quite possible; the only question is whether the Brothers are "thrice blessed" with the knowledge and divine grace they claim:

> Neither (since I know them not) did I ever aver constantly that they were so divinely possessed, or could do that they published, but did affirm in mine Apology that what they professed was possible to be accomplished, as well by natural as supernatural means.[32]

And as Scriptures confirm, there are elect sons of God on earth who do possess the hidden Manna, the heavenly bread, the spiritual rock, the divine sapience, the all in all, the Christ Spirit; and if the Brothers possess it, they can do all they claim:

> And verily if the Fraternity of the Rosy Cross in whose behalf mine Apology was written, for the which I have been without due reason or occasion suspected and brought in some question, If they (I say) have been so fortunate through grace to have lighted upon this lovely creature . . . as they themselves do with the sound as it were of a trumpet profess unto all the world, I make no doubt but all ye that shall perfectly understand so much will assert with me that these are to be numbered among the rank of those of which Solomon speaketh . . . [33]

Besides its revealing information, the "Philosophical Key" is also noteworthy for the fact that it was never published. The tract, "Of the Excellency of Wheat," was clearly intended for publication, because notes to the publisher on where to insert illustrations are included with the text, and so it is reasonable to assume that the entire manuscript, including the short Epilogue, was designed for the printer. In fact, the alchemical experiment on wheat did appear in print in the *Anatomiae Amphitheatrum* of 1623, and the title *Clavis Philosphiae et Alchymiae Fluddanae* was used ten years later in a polemical work against Mersenne and Gassendi, as was discussed in Chapter V; but the Dedication to James I, the Preface, the "Calumniators Vision" and the Epilogue disappeared from public view until 1963, and were only published in their entirety in 1979. The result was, with respect to the Rosicrucians, that Fludd was basically known for his Apologies, in which he had defended the possibility of their claims, and appended addresses hoping to attract the Fraternity; and by the curious work, *Tractatus Theologo-Philosophicus*, in which the certainty of the existence and spiritual nature of the Brotherhood was proclaimed in the boldest terms. Without the testimony of the "Philosophical Key," there was little to allay the notion that Fludd's supplications were rewarded by initiation into the august order.

This idea was reinforced by his presumed relationship with the other major Rosicrucian apologist, Michael Maier, the German physician and alchemist, about whom little is known. Maier was born in Rendsburg, Holstein, about 1568, and apparently received his doctorate in medicine from Rostock, where he also resided for a time. Sometime after 1608, he was summoned to Prague and became body physician to Emperor Rudolf II. During his residence there, the Emperor elevated Maier to the nobility by making him a Count Palatine and also appointed the German alchemist to be his private secretary. Since Rudolf gave up the crown in favor of his brother in 1611 and died the following year, Maier probably left the imperial service in those years. He apparently made a trip to England about 1614–16 to confer with like-minded alchemists, and it was there he first learned of the Rosicrucian manifestos. In 1619, Maier was the physician to Moritz, Landgrave of Hesse, but the following year was practicing in Magdeburg, where he died in 1622.[34]

While it is certainly possible and even reasonable to suppose that Maier and Fludd may have met in 1614–16 or subsequently, there is no direct evidence whatsoever of such a meeting, much less that the Count Palatine initiated Fludd into the Rosicrucian Fraternity. There are, however, sufficient points of commonality to warrant a presumption of their acquaintance, or, at the least, correspondence. About 1614 Maier published his *Arcana arcanissima*, which was dedicated to Sir William Paddy (1554–1634), physician to James I and also a graduate of St John's, Oxford (B.A., 1573).[35] Fludd also dedicated one of his own major works to the older English physician, the first volume of his *Medicina Catholica* of 1629, and

Paddy in turn donated two of Fludd's books to the library of St John's.[36] Paddy was also President of the College of Physicians in 1618, the same year that Fludd was a Censor of the College and defended his philosophy and Rosicrucian Apologies to James I. Paddy possessed a considerable number of books (many of which were donated to St John's) and frequented the great library of his good friend, Sir Robert Bruce Cotton, to whom Fludd also dedicated a book.[37] Maier very likely met Paddy on his trip to England, and certainly the German physician came to seek out such men of learning.

Another link between Maier and Fludd revolves around the latter's treatise on wheat in the "Philosophical Key." As pointed out by C. H. Josten, Michael Maier had mentioned a *Tractatus de Tritico* (*triticum* is the Latin for wheat) in his splendid emblematic book *Atalanta Fugiens* of 1618; and beside Fludd's entry in the commonplate book of Joachim Morsius, who made a trip to England from October 1619 to February, 1620, one of the English philosopher's books is listed as a *"Lib. de Tritico."*[38] It was noted above that the alchemical experiment on wheat was published in the *Anatomiae Amphitheatrum* of 1623, shorn of its other revealing material in the "Philosophical Key," and was entitled *"De exacta alimenti panis sev tritici anatomia."* Considering the importance of the wheat experiment for Fludd's work and that it probably existed in manuscript in 1618–20 or earlier, as well as the fact that no other work is known by that name, it appears most likely that they are all one and the same.

Over the years many writers have made much of the fact that Maier and Fludd clearly shared two things: they were the most prominent and eloquent defenders of the Rosicrucians, and they both had important works published by Johane Theodore de Bry of Oppenheim and Frankfurt in 1617 and 1618. In his *Silentium Post Clamores* (1617), Maier attempts to explain why all the clamor caused by the manifestos has been met with silence.[39] Here he undertakes a general defense of the order, which is on much the same lines as Fludd's. There have been many such orders (he says) through the ages which have possessed the same arcane wisdom and universal medicine; the manifestos contain nothing contrary to reason, experience and nature; therefore it is not incredible that this order now exists, and they are not guilty of necromancy, sorcery or evil. The silence of the order resulted from the unworthiness of many who wrote to seek its protection and afterward turned against the Fraternity. Furthermore, the ancient Egyptian initiates were pledged to secrecy, as symbolized by the Sphinx, and the Pythagoreans were to maintain silence for five years to keep the great secrets of Nature from those not worthy, and thus the Rosicrucians must also maintain the same period of quiet.

Another of Maier's works of 1617 was the *Symbola Aureae Mensae*, in which he expresses his lack of doubt about the truth of the manifestos, and attempts to reconcile their contradictions.[40] In the *Themis Aurea* of the following year, Maier expounds at length on the "Laws of the Fraternity of the Rosy Cross," which were the same six given in the *Fama*,[41] as listed above. With these three works, it is not difficult to see why Maier had the reputation of being a Rosicrucian initiate. Particularly through the *Themis Aurea*, he seemed to wish to cultivate the image of the Fraternity's real existence and his own knowledge of it, although he denied being a member.

The fact that Fludd and Maier both had publications produced by Johann Theodore de Bry does not appear to be accidental. As discussed earlier, in the "Declaratio Brevis," Fludd said that the person "to whom I entrusted this volume [the *History of the Macrocosm*] in England" wanted to dedicate it to his own prince, the Landgrave of Hesse. Thus whoever took the manuscript from London to de Bry in Oppenheim was a subject of that ruler, and considered worthy enough by Fludd to be entrusted with the valuable document. Maier dedicated three of his own books to the Landgrave: *De Circulo Physico* (1616); *Verum Inventum* (1619); and *Septima Philosophica* (1620). Since he was in the service of the Landgrave about that same period (c. 1616–1620), it is certainly well within the range of possibility that it was Maier who carried Fludd's manuscript to Germany. This is further strengthened by the fact that Maier must have had a number of manuscripts of his own to hand, because his considerable number of works were almost all published between 1616 and 1621; ten of them appeared in 1617 and 1618, all from the presses of Lucas Jennis and Johann Theodore de Bry.[42] Fludd had said that the English printers wanted him to put up five hundred pounds and pay for his own copper cuts to have the first volume printed, but that de Bry published his *History* at no cost to him, and even sent sixteen copies and forty gold pounds as a gratuity.[43] It may well have been Maier who found a compatible publishing house for his own as well as for Fludd's manuscripts, which had been so long in the making.

It is reasonable to speculate that Fludd and Maier knew each other and that their shared world-view, which included agreement on the possibility of a Rosicrucian Fraternity, resulted in their works being printed by the same publisher. One could also imagine that for some reason (perhaps political, as Dame Frances Yates has suggested) Maier wanted to promote the idea of the Fraternity; and that he read Fludd's "Philosophical Key," and decided that its self-deprecating tone and despair at ever hearing from the Fraternity would discourage that idea. Perhaps he therefore persuaded Fludd not to have it printed, and indeed it never was. It is odd that Fludd's experiment with wheat, which he says was the foundation of his entire philosophy and which apparently existed as a manuscript as early as 1618, was not printed until 1623, one year after Maier's death. The possibility could also be entertained that the two learned physicians kept their conferences and correspondence secret, lest they be accused of being members of the Fraternity. Maier's purposes seemed to have been better served by proclaiming the reality of the Fraternity, but only from a distance. The very fact that Fludd and Maier never mention each other directly, considering their closely held beliefs, supports the idea of a privately-kept relationship. Their presumed connections do not add any evidence whatever for their mutual membership in the Fraternity of the Rosy Cross, however.

In fact, everyone's experience in Europe during this period was the same: the Brothers remained invisible and silent, much to the great disappointment or anger of many. Whether for or against, many at first believed in their existence; as time passed with no contact, some came to think it was a hoax, and others, who did not wish to stop believing in them, simply allowed the order to remain in a spiritually ideal state where they chose not to appear to others at this time. Fludd fell in the latter category. Rumors of their existence died hard, however, and they were kept alive by periodic reports of sightings, or supposed missives issued by the Brothers, and the like. It is interesting to note, for example, that René Descartes spent the winter of 1619 in Germany looking in vain for the Brothers, and returned to Paris in 1623 at the height of a "Rosicrucian scare" in the city. It seems that placards were placed in the city announcing the arrival of the College R. C., Marin Mersenne, Fludd's adversary, rushed to see the newly arrived Descartes, and was relieved to find out he was not a member of the order, as had been rumored.[44] As usual, nothing came of the affair beside a great deal of excitement and speculation.

From Fludd's point of view, he remained steadfast in the principles and ideals of the Fraternity, and indeed for him they remained at that lofty level. In fact, after his Apologies of 1616–17 and the unpublished manuscripts of 1618–19, he has nothing more to say about them until 1633. There was, however, a publication of 1629 which again many used to tie Fludd to the order. In that year he published a reply to an attack by Mersenne called *Sophie cum Moria Certamen*, which does not deal with the Rosicrucian issue. Bound with it, though, was a small work of fifty-three pages which did, bearing the title *Summum Bonum*, by Joachim Frizius. Since this tract also is a reply to Mersenne, the printers (Officina Bryana, Frankfurt), in a note to the reader, say they decided to bind it with Fludd's work on the same topic.[45] The *Summum Bonum* takes a highly spiritual, mystical approach to the subjects of Magic, Cabala, Alchemy and the Fraternity of the Rosy Cross, and backs up its arguments with many references to Scripture and compatible authors such as Pico, Ficino and Reuchlin. At the end of the *Summum Bonum*, Frizius appends a letter supposedly written by the Fraternity to a German candidate who has just completed his first year in the order. The letter, he said, was obtained by a Polish friend in Danzig, and purports to show both the existence of the Fraternity and their high Christian faith and philosophical ideals.[46] Since the style, subject matter and publishing history make it circumstantially seem to be a Fluddean work, it has been assumed to be so, particularly by those who wish to have him be a member of the Fraternity.

Fludd specifically denies authorship of the *Summum Bonum* in his *Clavis Philosophiae et Alchemiae Fluddanae* of 1633: "It is not mine, but the work of an intimate friend."[47] He also said that he wanted the printer to produce it separately, in octavo, not in folio with his own reply to Mersenne.[48] Perhaps he feared exactly that which happened, that the tract would be thought his, and he in turn be believed to be a Rosicrucian. It shows as well that Fludd knew of the *Summum Bonum* and probably gave it to the printer, perhaps after editing, and he clearly says that an attack on Frizius was also an attack on his own ideas.[49] The end result was another connection presumed between Fludd and the Fraternity without any real justification.

Some sixteen years passed between the publication of his Rosicrucian Apologies and his last (and only other) printed words on the subject. The *Clavis* of 1633 was a detailed and final response to the attacks on his philosophy by Mersenne, Lanovius and Gassendi, and defends it with the same basic arguments as in his other works.[50] Here the Rosicrucians and their work are totally framed in the language of Christian alchemical allegory. In agreement with Frizius' statement in the Epilogue of the *Summum Bonum*, Fludd repeats that the true Brothers of the Rosy Cross are those called the Wise, the *theologi* of the Church Mystical. They are among those in all religions, whether Papal Luterhan, Calvinist or others, and are there regardless of their adherence to external rituals and ceremonies. Such seekers of divine wisdom are thoroughly grounded upon the spiritual rock of Christ, and thus they disdain the common gold and silver, and work for the spiritual transformation which results in the finding of the true Philosopher's Stone, or Christ-spirit within.[51]

In repeating Frizius' statements of 1629, Fludd emphasizes the allegorical side of the *Fama* and *Confessio*, which is probably the only reasonable position left to take, since in all the intervening years the Brothers apparently were not genuinely heard from by anyone. It is also interesting to note that, in contrast to the original manifestos, which were violently anti-papal, Fludd and Frizius would have members of the order to be found in all Christian branches. From the point of view of the *Summum Bonum* and the *Clavis*, Fludd and Frizius would seem to qualify as candidates for such an allegorized Rosicrucianism, although both would deny being worthy of membership. But Fludd's belief in his possible candidature ended about 1619–20, as he reported in the "Philosophical Key," and he never publicly commented on the issue until the *Clavis* of 1633, when he was answering the attacks of Lanovius, Mersenne and Gassendi point for point. Even then, most of the comments revolved around what was said in the *Summum Bonum* and whether it was Fludd's work. In 1629, he could have replied on the topic in the *Sophie cum Moria Certamen*, but did not do so, and instead alllowed the Frizius tract to carry the burden, and in the *Clavis* he basically defers to Frizius' points.

In his two other books of the 1630s, *Dr. Fludd's Answer unto M. Foster* (1631) and the *Mosaicall Philosophy* (1638, 1659), he also declines to mention the Brotherhood. On p. 23 of the former he simply says, when Foster repeats Lanovius' assertion that the *Summum Bonum* was Fludd's work, that he has satisfied Gassendi on the matter, whom he esteems far greater than Lanovius. Fludd says that he does not care to take up the issue "at this time," but to the charge that the *Summum Bonum* excuses Roger Bacon, Trithemius, Cornelius Agrippa, Marsilio Ficino and the Fratres Roseae Crucis from being cacomagicious, he adds:

> . . . onely thus much will I say for Joachimus Frizius, that what he hath produced out of their own workes, in their own defence, excuseth them, and accuseth such calumniators as Master Foster is, who are so apt to condemne a person for that they are altogether ignorant in: Let the Readers observe the proofes in Frizius his booke to cleare them; and then if any will afterwards excuse them, I shall deem them partiall.[52]

Here again he simply defers to Frizius, and in his last, posthumously published book, the *Mosaicall Philosophy*, he does not discuss the issue at all, even though the book covers the entire range of his philosophy and takes on his critics at large.

Thus, considering our original questions in reverse order: firstly, was Robert Fludd a Rosicrucian? In the sense of belonging to an order of the Brothers of the Rosy Cross announced in the *Fama* and *Confessio* of 1614 and 1615, the evidence says he was not. It would certainly make a better and more interesting history of the Rosicrucian affair if the case could be made for the membership of both Maier and Fludd in the Fraternity, at least from the stand-point of an actual operating organization, but their experience was exactly that of the others trying to contact the Brotherhood: "*silentium post clamores.*" Some five or six years after the appearance of the manifestos, Fludd swears that he knows nothing of them, even though three or four years have passed since he published his apologies in their favor with attached letters of supplication. In the "Philosophical Key," he calls his Apology "silly and poor," and seems sufficiently embarrassed by the affair to never truly deal with the issue at length again, except to say that he agrees with Joachim Frizius' *Summum Bonum*; and Frizius' tract has so totally fashioned the Brotherhood to be part of a high Christian-alchemical mystical system that their relationship with the original manifestos is almost unrecognizable. Fludd's own words clearly support the conclusion that he was not a member of the order.

Would it have been possible for Fludd and perhaps Maier to have belonged to the order and kept a vow of silence, and this is why their membership has not come to light? Working against this idea are two things. First, if an actual order existed, it has remained one of the best kept secrets in history, which seems unusual in the normal run of human affairs, but particularly so when so many were eagerly searching for some contact with or knowledge of it, as have subsequent generations. Second, Fludd's membership would have to have been post-1620 or 1621, when he wrote the "Philosophical Key." As Fludd himself lamented, too much time had passed since the publication of the manifestos and his Apologies for him to expect contact any longer. Of course, there was a major reason why he probably never would come in touch with them after that date, even if they existed as a working order: the prosecution of the Thirty Years' War on the Continent, particularly in Germany. Once again, it is difficult, if not impossible to even build a good circumstantial case for Fludd's membership in an actual organization as specified in the manifestos.

Fludd's reputation for being a Rosicrucian rested upon the weight of his Apologies, his presumed acquaintance with Maier, and his assumed authorship of the *Summum Bonum*. The authors quoted at the beginning of the chapter, except Frances Yates, did not have access to the "Philosophical Key", and all were willing to believe in the existence of an actual Fraternity of the Rosy Cross. Thus until the mid-1960s, when this revealing manuscript came to light, it was possible to speculate about Fludd's membership with some degree of credibility, but it is certainly difficult to maintain that view today. Indeed, the only remaining question about Fludd being a Rosicrucian is in the special sense used by Dame Frances Yates. She employs the word in a historically cultural sense, to stand for the Renaissance Christian Hermetic-Cabalist-Alchemical school of thought which ended in the seventeenth century, of which Fludd, Dee, Maier and the manifestos are the leading examples. This Rosicrucian cultural movement she also attempts to link with Protestant power-politics on the Continent just prior to the Thirty Years' War.[53]

While there is certainly some justification for calling Fludd a

Rosicrucian from this point of view, and no doubt that this particular intellectual movement is in need of a good, universally accepted term, the word "Rosicrucian" in this context brings some difficulties with it. One is, of course, that historically, as demonstrated earlier, Fludd was disdained because he was assumed to be a Rosicrucian and therefore was a fringe occultist not to be taken seriously. Another is that use of the term implies membership in a secret society of that name, reasonable proof of which has yet to surface. A further important objection is that Fludd himself would surely not wish to have his philosophy known by that name. Allen Debus has included Fludd as major figure of the "chemical philosphers," a term used at the time, and it is quite appropriate, particularly from the point of view of the history of science. However, Fludd, who was put off by the Rosicrucian non-appearance but did advocate a "true chemical philosophy," chose to name his work "Fluddean," and, finally, the "Mosaicall Philosophy." I believe this latter term reflects better the all-important mystical Christian underpinnings for all of Fludd's thought. Last of all, Rosicrucianism implies some sort of homogeneous movement, within or without a formal organization, in which, through consensus, the adherents share a common ideology and goals, perhaps including political ones, as Dame Frances suggests. But, as Robert Westman has clearly shown, among the Hermetic thinkers of this era there was great diversity in both method and conclusions, and followers of Neoplatonism and Hermes could even reach opposing results, such as those of Fludd and Kepler.[54] And in this same way, "Rosicrucianism" cannot be used to clearly identify a homogeneous group with direct links to one another, as, for example, between Fludd and John Dee or Michael Maier, even though they may incidentally share a broadly common world-view, because the evidence simply does not exist to support it. Therefore, calling Robert Fludd a Rosicrucian, even in the broader sense, is probably more misleading than helpful as a description. The same would be the case for other figures of the time.

The question of whether there was a formal, operating Fraternity of the Rosy Cross may never be known with certainty, but this is not as important an issue as the broader perspective of the Renaissance ideals expressed in the manifestos and the responses to them. From the best assessment to date, it appears that the origins of the idea for a Rosy Cross Fraternity are to be found in the late sixteenth century, and that the manifestos themselves may have circulated in manuscript in Germany as early as the 1590s, and perhaps earlier.[55] They could have issued from a circle of German Protestant mystic enthusiasts connected with Simon Studion (c. 1543–16?) of Wurtemberg, whose long, unpublished manuscript "Naometria" (written 1593–1604) is apparently closely related to the Rosicrucian manifestos.[56] As Montgomery has pointed out,

> Political and social conditions were sufficiently bad at this time that chiliastic mystics of Studion's brand came to believe that a second reformation – a "general reformation" – was about to come which would fulfill the frustrated, millennial expectations they had had for Luther's reformation.[57]

In order to help bring about this new age, they may have wished to gather together the like-minded for further instruction in the ancient wisdom and to foster cooperation among them. For this purpose the manifestos were probably written, and circulated in manuscript for perhaps twenty years or more.

For what reason the *Fama* and *Confessio* were finally published in Hesse-Cassel under Landgrave Mortiz is not clear. It may have been an attempt, as Dame Frances Yates suggests in *The Rosicrucian Enlightenment*, to establish an ideological base for the Princes of the Protestant League on the Continent, with Frederick of the Palatinate at its head, a movement which ended tragically with Frederick being deposed as King of Bohemia and the invasion of the Palatinate at the beginning of the Thirty Years' War. They may, however, have been meant to serve a somewhat different, less directly political purpose. Perhaps the intent was to bring about the "general reformation" through building a cadre of fellow-believers in the Christian-Cabalistic esoteric wisdom, an order akin but opposed to the Jesuits, which would in turn attempt to convert the learned and the political leaders of Europe. If ultimately successful, it would result in thoroughly reforming Western civilization philosophically and religiously, from which could follow a politically reunited Holy Roman Empire. Thus would the millennium, a new golden age, ensue. This, in essence, was the Renaissance ideal as a whole.

In its first edition, the *Fama* was preceded by the Boccalini satirical piece, "Universal and General Reformation of the Whole

Wide World," which despairs of ever achieving reformation by relying on the learned establishment. Not only are they unqualified for the task, they always attempt to hide their incompetence and thereby save their reputations by loudly trumpeting mere trifles as things of profound importance, so that the real reformation is continually left unaccomplished. As Apollo knew, "men are better improved by the exemplary life of their reformers than by the best rules that can be given." This was followed by the *Fama* itself, which repeats his theme by showing that Brother R.C. and Paracelsus were equally unsuccessful in reforming learning in Europe according to their superior understanding of nature because the established scholars feared a loss of prestige and position if they admitted their errors of many years' standing. To surmount this obstacle, the Fraternity was formed, and each Brother went to a different country to quietly bring the deserving intelligentsia and whatever rulers they could, present and future, into the fold. Since the discovery of the tomb of Brother R. C. heralds the time to push forward with the great reformation on a larger scale, it becomes appropriate to publicly reveal the Fraternity and call for the like-minded to join them. The *Confessio* continue the same idea in a more militantly apocalyptic fashion.

As expressions of the Renaissance ideals of reunion and reform based on lost ancient wisdom, the manifestos are quite understandable. It is not difficult to imagine some German Protestant mystics of the late sixteenth century who were dissatisfied with the incomplete state of reform in both learning and religion, which hindered the resulting unity and thus the coming of the millennium. They could have conceived the myth of the Brothers of the Rosy Cross as a foundation for establishing a real brotherhood, to be the Protestant equivalent of the Jesuits, to seek followers for their cause at princely courts and universities. For some reason, perhaps because of suppression by influential clergymen or noble patrons who feared the political consequences or objected to the unabashed alchemical occultism of the documents, or both, the manifestos remained in manuscript for twenty years or more, and the Fraternity itself remained very small, or was not sustained as a formal organization, or was never constituted in the first place.

The *Fama* and *Confessio* were finally published, along with Boccalini's satire and the John Dee-based "Brief Consideration", in 1614 and the following year in Cassel. By that time and in that place some remnant of a Brotherhood R. C. may or may not have survived, but in any case it could not have been large or influential. The manifestos and accompanying tracts were, however, published in a principality (Hesse-Cassel) whose rulers had a long history of deep involvement in alchemy as well as in other Renaissance arts and sciences. Both Landgraf Wilhelm IV, "the Wise" (r. 1567–92), and Moritz, "the Learned" (r. 1592–1627), were personally acquainted with John Dee, who visited their court in Cassel in 1584 or 1585, and with whom they carried on a correspondence.[58] Their court was known as a center for a broad range of artistic and scientific endeavor, which included Renaissance esoteric and alchemical studies.[59] Also, the province became Lutheran in 1526, and in 1605 Calvinism was introduced under Moritz. Landgraf Wilhelm had allied himself with Moritz of Saxony in the religious wars against Emperor Charles V in the mid-1500s. Thus the atmosphere was certainly warm for the publication there of the manifestos and the other material. After their appearance, it was the Landgrave's personal physician, Michael Maier, who attempted to keep the myth alive by his writings.

It would seem that bringing out the Rosicrucian tracts from the press of Wilhelm Wessel, whom Moritz had set up in business in Cassel in 1596 as the court printer, had a purpose beyond perpetrating a deliberate hoax to stir up some excitement. Landgraf Moritz, like his father before him, was a serious student of theater, poetry, philosophy, philology, medicine, botany, astronomy and chemistry; in short, a true Renaissance "universal man" in the style of the Medicis. Just as Michael Maier's other noble patron, Emperor Rudolf II, had done, Moritz turned over the reins of government to his son, Wilhelm V, in 1627, in order to retire to his studies for the rest of his days.[60] If the Landgraf knew and approved of, or even decreed, the publication of the *Fama* and *Confessio*, which seems likely, it would probably have been done with the collaboration of some of his courtiers, most notably Michael Maier, and had some serious intent.

While it is possible that they still wanted to establish a Rosy Cross Fraternity to carry out the utopian aims of the originators of the manifestos or the political aims of the Protestant League, it is also quite possible that they simply wished to set up some practical means of cooperating among like-minded Hermetic-alchemical scholars in Europe for the betterment of all concerned. The lack of such cooperation was clearly a major complaint in the *Fama*, and one that Moritz and Maier must have experienced many times. Prior to the founding of the Royal Society in England and similar societies on the Continent, this was a commonly perceived problem. The most famous voice calling for institutional cooperation among scholars in the early seventeenth century, of course, was Francis Bacon, whose hopes were only realized after his death with the etablishment of the Royal Society in 1661.

It may have been the case that the Rosicruciana were published to help identify the prospects for a cooperative enterprise, but the volume of response was completely unanticipated. It could have been thought that the right people would discern their intent through the myth, allegory and symbolism of the two books, but the ensuing uproar and confused proliferation of replies caused them to abandon the project or hold it in abeyance. Certainly Andreae's intent in writing the *Chemical Wedding* was totally lost by being caught up in the Rosicrucian mania and considered to be a work issuing from the inner circle of the Fraternity, when in fact it was written as alternative to it. Maier's books on the Rosicrucians seem to have been designed to keep the notion of the Fraternity going in the aftermath of so much disillusionment, caused by the failure of the Brothers to contact anyone, but forces had been set in motion which were mightier than the wished-for reformation of the manifestos.

In the end, whatever ambitions for a philosophical and scientific enterprise based on Renaissance ideals were held in the late sixteenth century and early seventeenth were doomed to failure. Time was making its inevitable turn on the spiral, and the world of the Renaissance was changing to that of the modern. The wisdom of the ancients soon would no longer be able to compete with the new ways of thinking based on a model of mathematical description divorced from religious metaphysics. The immediate event that stifled intellectual life in Germany and prevented any eleventh-hour postponement of the demise of the traditional modes of Renaissance thought was the all-too-usual one for that country, war. With the outbreak of the Thirty Years' War, the German phase of which started with the depositon of Frederick of the Palatinate as King of Bohemia in 1620 by Emperor Ferdinand, and the Spanish invasion of his Heidelberg-based principality that same year, all hopes of a grand collaboration of Protestant German Hermetic-alchemical scholars with those of other lands were buried forever. On the other side of the watershed, when the Thirty Years' War and the English Civil War had run their course, collaboration indeed began with the establishment of scientific societies, formal and informal, in Italy, France and England, but of course this hastened the burial of Rosicrucian-type formulas for changing the world in favor of the modern one of faith in empirical, mechanistic science to lead us to the millennium.

Judging by the great number of responses to the Rosicrucian appeals, there was much sentiment in favor of the ideals expressed in them. Had they been an operating order taking an active part in the intellectual life of the day, however, they would have suffered the same fate as that of Robert Fludd: constant warfare with the mathematicians and mechanists, resulting in eventual relegation to the role of an unimportant curiosity of history and an example of how absurd the thought of well-known pre-modern figures could be, compared to modern concepts. Today the Rosicrucian phenomenon remains a matter of interest because of the mystery surrounding the originators of the manifestos, the undeciphered mythological content, and the possible conspiratorial intent, which did not succeed. But viewed in a larger context, it may be seen as one of the final expressions of the desire of the Renaissance to complete the utilization of ancient esoteric wisdom for reform of the arts, sciences and religion, a process which would return society to the idyllic unity lost in the remote past. Thus did Robert Fludd and the writers of the Rosicrucian manifestos share the same fortune, though they did not know each other, by living at the end of their era.

Chapter X

DE VITA, DE MORTE, DE RESURRECTIONE

> Although I be never so curious in my inditing, or laborious in
> the phrase of mine expression; yet will My best endeavors
> appear faulty in the curious eyesight of some men, though
> perchance acceptable enough for others. I esteem it
> sufficient . . . that I dare be hardy and bold in the essentiall
> Philosophy, being that it hath Truth itself to maintain and
> defend it.
>
> Fludd, *Mosaicall Philosophy*

In 1630, a collection of portraits and short biographies of illustrious
men appeared in Frankfurt. It said of "Robertvs Flvdd,"

> We place here the inscription on the earlier portrait of Robert
> Fludd, who is still living, and who shared the splendor of his
> name not only throughout Britain but also foreign kingdoms. For
> indeed honorable is he who . . . is descended from the ancient
> and noble line of Fludds . . . and honored with the title of
> Esquire . . . aspiring to the true and long-lasting glory which
> promises immortality from his writings, he decided to investigate
> the inmost secrets of Nature herself, the great Mother, in order
> that he might add a rich chorus to the art of Medicine in which
> he had attained the greatest ranks of honor as well as glory.
> Almost no one from the entire gallery of men of letters failed to
> give applause to his works . . . [1]

Despite this high praise, by the end of the century Fludd's works
were little read, and for the next two-and-a-half centuries he
became almost completely unknown. A major feature of Fludd's
place in history is the change from a position of prominence,
respect and modest fame in his own time to one of relative
obscurity until recently. In the last few years, several major scholars
have shown the importance of Fludd's works for a more complete
understanding of this crucial period of transition from traditional
to modern thought. Their pioneering work has inspired a greatly
renewed interest in Fludd, including my own. This book is

intended to be a part of the growing modern literature about Fludd that should re-establish him as a significant figure of his time.

Fludd's historical esteem did not fall because his religious-Hermetic-Neoplatonist metaphysics were exposed as false by the relentless forward march of reason through time; it declined because of a change in the commonly accepted fashion about *sources* of knowledge and the consequent *methods* used to achieve it. It is the nature of scientific revolutions that they do not follow a steady course of the revelation of truth over time in a linear manner; a new paradigm completely replaces the old by a process of total negation. In turn, the new paradigm eventually will be unable to accommodate non-conforming new information, and will go the way of its predecessor. No rationally-based paradigm can ever reach the status of absolute truth, because of the very nature of its incompleteness.[2]

Robert Fludd's grand and elaborate synthesis was accomplished just as other voices were signaling new directions for science and philosophy. Francis Bacon was calling for a new science not based on religion or the ancients. Descartes was unintentionally solidifying doubt about all metaphysical knowledge. Kepler was insisting on the mathematically demonstrable geometry of the heliocentric universe, and Galileo was destroying ancient concepts with his telescope and his mathematical calculations.

In the early seventeenth century, however, Fludd's voice was just as strong as, and often stronger than, the distinguished company just cited. His work was the greatest, but last, expression on that scale of the concepts revived and added to by Marsilio Ficino and Giovanni Pico della Mirandola in Renaissance Florence. As so often happens in history, philosophical concepts reach the peak of their expression just when they are near their end, following which they are replaced by their opposite. It is the process of *enantiodromia*, where values reverse through time, which has been described so well by William Irwin Thompson.[3] Thus Plato is followed by Aristotle, and Fludd's Neoplatonic universe, powered by the divine spirit, is followed by Newton's clockwork, run by gravity.

After the mid-century, following the Thirty Years' War on the Continent and the Civil War in England, the new science became institutionalized by learned societies with royal backing. By the end of the century, the works of John Locke and Isaac Newton appeared; their methods were propagated by the Enlightenment figures of the following century, and eventually applied to all areas of thought. In such a framework, it is hardly surprising that Fludd, who looked back to the ancient metaphysical wisdom of Plato, Hermes and the Bible as his mainstays and incorporated it in long, complex tomes which completed the work of the Florentine Academy of Ficino, no longer received acclaim from the learned community, but was indeed relegated to the position of an unimportant curiosity. But such a selective view of history creates an unacceptable distortion of historical fact, which prevents a true understanding of the time. Every age considers the previous era to be one of ignorance and folly, and selects out the works of certain figures to venerate which later proved to be forward-looking or which contained qualities of special value to later generations. Often the work, or the person, or both, are taken out of their contemporary context, and other ideas of these figures, which do not fit modern views, are actively ignored or credited to the superstitious ignorance of the times. Certainly this was, and still very often is, true of Francis Bacon, as it is of Isaac Newton and other figures of the time. Those who did not have important features of their work that were compatible with the new scientific age were often literally written out of history. In Fludd's case, it was the valuable as well as the "modern" parts of his work, and his substantial place in the intellectual and professional communities, which were forgotten as he became known only for curious occult metaphysics, being a "Rosicrucian," and having an odd medical practice.

Apart from the controversies with Kepler, the French mechanists and William Foster, Fludd's reputation remained high until the mid-seventeenth century, although the seeds of destruction were planted during his lifetime. One of the difficulties Fludd's legacy ran into was the link that others made between him and two Elizabeth practitioners of the occult arts, the famous John Dee (1527–1608)[4] and the lesser known figure, Simon Forman (1552–1612).[5] Fludd never mentioned either of them in his own writings (quite possibly because of his awareness of their reputations, and his own disdain for the darker side of the occult), but this did not prevent him from being associated with them by others in a negative way in the seventeenth century and later.

Queen Elizabeth greatly admired Dee's learning and skill: she had him cast a horoscope to choose the best time for her coronation to ensure a long and prosperous reign, and she followed it to the letter.[6] The Queen and her entire court made a special trip to Dee's house in Mortlake to see his library, by far the largest in England at the time.[7] In her characteristic fashion, however, she assured Dee of "suitable" preferments, but delivered little he found suitable. Dee became quite bitter about the state the Queen had allowed him to fall into, which forced him to sell precious books from his library in order to eat. There can be little doubt that the Queen and her advisors felt it necessary to maintain some distance from Dee because of the reputation he had acquired as a conjurer and black magician, particularly through his experiments with his "medium," Edward Kelley.

On James I's ascension to the throne, Dee even demanded a trial on the slanderous charges making the rounds about him, in order to clear his name once and for all, but the crown discreetly ignored his pleas. Meric Casaubon's publication in 1659 of Dee's record of some spirit-summoning sessions succeeded in blackening Dee's name for the next two-and-a-half centuries, and tied Fludd in with him:

> The Divil is not ambitious to shew himself and his abilities before man, but his way is (so observed many) to fit himself (for matter and words) to the genius and capacity of those that he dealeth with. Dr. Dee, of himself, long before any Apparition, was a Cabalistical man up to the ears, as I may say; as may appear to any man by his *Monas Hieroglyphica*, a book much valued by himself, and by him Dedicated at the first to Maximilian the Emperor, and since presented (as here related by himself) to Rudolphe as a choice piece. It may be thought so by those who esteem such books as Dr. Floid, Dr. Alabaster, and of late Gafarell, and the like.[8]

Thus Dee's work was branded as diabolical magic, and Fludd, no longer alive to defend himself, depicted as the same sort of person as Dee.

Simon Forman, a self-taught astrological medical practitioner and seer who was arrested on a number of occasions by the College of Physicians for practicing medicine without a license, enjoyed a much more unsavory reputation than Dee. Forman had a large and lucrative practice among many of the high nobility of England, and was ultimately granted a license to practice medicine from Cambridge, although he had not formally studied there. The main cause of the ruin of Forman's reputation came from the false linking of his name with the poisoning of Sir Thomas Overbury by the Countess of Essex in 1613, when allegations were made that the Countess had consulted Forman for her scheme. Even though Forman had died two years earlier and had nothing to do with the scandal, some of those involved had consulted Forman in the past, and his name was repeatedly invoked at the trial and vilified by Lord Chief Justice Coke.[9]

Fludd was associated with Forman's manuscripts by two seventeenth-century sources. William Lilly, the renowned astrologer and astrological writer in the mid-century, wrote a *History of His Life and Times* (1667) for his admirer Elias Ashmole, wherein he says of Forman:

He was a person of indefatigable pains. I have seen sometimes half one sheet of paper wrote of his judgment upon one question; in writing whereof he used much tautology, as you may see yourself (most excellent Esquire) if you read a great book of Dr. Flood's, which you have, who had all that book from the manuscripts of Forman; for I have seen the same word for word in an English manuscript formerly belonging to Dr. Willoughby of Gloucestershire.[10]

This passage was picked up by Anthony Wood when he wrote his still-used *Athenae Oxonienses*, a collection of biographies of Oxford-educated writers and bishops. In the article on Forman he states, "he used much tautology, as you may see if you'll read a great book (*in musaeo Ashmoleano*) of Dr. Rob. Fludd, who had it all from the MSS. of Forman."[11] Whether or not there is any truth in this assertion remains to be investigated, but the fact remains that Fludd has been associated with a thoroughly discredited Simon Forman for over three centuries.

One of the major forces working against Fludd indirectly was the opinion of the one-time Lord Chancellor Sir Francis Bacon, Baron Verulam, Viscount St Albans. Although he never mentions Fludd by name, there can be little doubt that he knew the nature of Fludd's work. One possible connection comes from Bacon's *New Organon* of 1620, wherein he discusses a "calendar glass" or "air thermoscope," which Fludd had first described in print in 1617, claiming that he (Fludd) had re-invented the device from a description in a five-hundred-year-old manuscript.[12] That Bacon disapproved of the main tenets of Fludd's philosophy is easy to show: when discussing the physics of light, for example, he says:

> Again the manner in which Light and its causes are handled in Physics is somewhat superstitious, as if it were a thing half way between things divine and things natural; insomuch that some of the Platonists have made it older than matter itself; asserting upon a most vain notion that when space was spread forth it was filled first with light, and afterwards with body; whereas the Holy Scriptures distinctly state that there was a dark mass of heaven and earth before light was created.[13]

Bacon is criticizing here exactly the view held by Fludd, and even using against him Fludd's mainstay, Holy Scriptures.[14]

A matter of principle for Bacon is his insistence that the Scriptures are not to be used as a basis for natural philosophy any more than the useless and even dangerous thought emanating from Pythagoras and Plato:

> . . . the corruption of philosophy by superstition and an admixture of theology is . . . widely spread, and does the greatest harm, whether to entire systems or to their parts . . . For the contentious and sophistical kind of philosophy ensnares the understanding . . .
>
> Of this kind we have among the Greeks a striking example in Pythagoras . . . another in Plato and his school, more dangerous and subtle . . . For nothing is so mischievous as the apotheosis of error; and it is a very plague of the understanding for vanity to become the object of veneration. Yet in this vanity some of the moderns have with extreme levity indulged so far as to attempt to found a system of natural philosophy on the first chapter of Genesis, on the book of Job, and other parts of the sacred writings; seeking for the dead among the living: which also makes the inhibition and repression of it the more important, because from this unwholesome mixture of things human and divine there arises not only a fantastic philosophy but also an heretical religion. Very meet it is therefore that we be sober-minded, and give to faith that only which is faith's.[15]

One of the "moderns" who indulged in a natural philosophy based on Genesis and many other passages in the Bible was, of course, Robert Fludd.[16] Indeed, a perusal of the Aphorisms in Book One of Bacon's *New Organon* makes one quickly aware that there are few which could not be directed against Fludd, such as:

> The study of nature with a view to works is engaged in by the mechanic, the mathematician, the physician, the alchemist, and the magician; but by all (as things now are) with slight endeavor and scanty success.[17]

Ironically, it was Bacon who was disgraced in 1621 and died without office, heavily in debt, five years later, while Fludd continued to enjoy royal favor, distinction and modest wealth until his death in 1637; but it was Bacon's views which survived and were hailed by subsequent generations, not Fludd's. The former Lord Chancellor has ever been looked back to for his "modern" outlook, particularly since he was elevated to near sainthood by many eighteenth-century writers.[18] There is no doubt that his ideas existed as a powerful force against those of Fludd, and that their success among the rising generation contributed a great deal to Fludd's passage to obscurity.

Although Fludd's polemical controversies with Mersenne, Gassendi and Kepler were conducted in Latin and meant for the learned only, they passed into the popular anti-Fludd lore, with Kepler's being the most important. Since Kepler came to be regarded a key figure in the "scientific revolution" on account of his correct theories of planetary motion, his own involvement in astrology and his Neoplatonic-mystical view of the universe were often ignored, and his modern outlook proved by, among other things, his attack on Fludd. The argument between the two was, in fact, the question of whose Neoplatonic structure of the heavens was the correct one.[19]

Parson Foster's attack on Fludd in 1631[20] dredged up some of the same objections Mersenne had used, and his book often took on characteristics of hysteria rather than logically tight argument. Fludd's *Answer unto M. Foster* deals rather decisively with the parson's contentions, and there can be no question that an unbiased reading of the two works reveals that Fludd successfully refutes every one of Foster's points. Both books were written in English – Fludd's because Foster's had been. Thus, both were available to a popular readership, which was certainly not always unbiased, and Parson Foster added to the body of literature, this time in English, against Fludd.

It is important for the purposes of this chapter to look at what was written about Fludd shortly after his death. Dr. Baldwyn Hamey the younger (1600–1676), a Fellow and benefactor of the Royal College of Physicians, compiled contemporary medical biographies in a manuscript which must have circulated in London. There is an entry for Fludd as follows:

> Dr. Fludd. Fellow of the College, lived splendidly enough, and died Septemb. 8th 1637. He continually supported, outside the custom of his colleagues, an amanuensis and apothecary at his house; the latter mixed and distributed medicines by day, the former received ideas that he had at night; in both of [these endeavors] he kindled not a little envy of himself; moreover, by his night studies, which was his custom to profusely produce, he seemed to undertake more work than our common people wished to enjoy; they mostly overlooked him because of the tediousness of reading him, and their prejudice against wasting time and oil, and because of the Cabalistic, rather than Peripatetic nature his writings are said to smack of, and because of the rather fervent character of the man, in whom many failed to find judiciousness.[21]

Sometime between 1648 and 1661, Thomas Fuller (1608–1661), a Doctor of Divinity by letter from Charles II in 1660, wrote the following article on Fludd in his well-known *The Worthies of England*:

> Robert Floid, who by himself is Latined *Robertus de Fluctibus*, was born in this County [Kent], and that of a Knightly Family, as I am informed; bred (as I take it) in Oxford, and beyond the Seas. A deep Philosopher, and great Physician, who at least fixed his habitation in Fan-Church Street, London. He was of the Order of the *Rosa-Crucians*, and I must confesse myself ignorant of the first Founder and sanctions thereof. Perchance none know it but those that are of it. Sure I am, that a Rose is the sweetest of Flowers, and a Cross accounted the sacredest of forms or figures, so that much of eminency must be imported in their composition.
>
> His Books written in Latine are great, many, and mystical. The last some impute to his Charity, clouding his high matter

with dark language, lest otherwise the lustre thereof should dazle the understanding of the Reader. The same phrases he used to his Patients; and, seeing conceit is very contributive to the well working of Physick, their Fancy, or Faith natural, was much advanced by his elevated expressions.

His Works are for the English to *sleight* or *admire*, for French and Forraigners to *understand* and *use*: not that I account them more judicious than our own Countrymen, but more inquiring into such difficulties. The truth is, here at home his Books are beheld not so good as Chrystal, which (some say) are prized as precious pearls beyond the Seas. But I conclude all with the Character which my worthy (though concealed) Friend thus wrote upon him: [here Fuller inserts the portion of Hamey's entry above beginning with "moreover"]. He died on the eighth of September, anno Domini 1637.[22]

During this period Fludd also had a devotee in the person of Elias Ashmole (1617–1692), the influential antiquary who left his considerable library to Oxford University. In his *Theatrum Chemicum Britannicum* of 1652, Ashmole laments the fact that some of Britain's most illustrious medical figures are ignored in their own country:

> . . . the Phisitians Colledge of London doth at this day nourish most noble and able Sons of Art, no way wanting in the choycest of Learning; And though we doe not, yet the World abroad has taken notice of sundry learned Fellowes of that Societie, as Linacres, Gilbert, Ridley, Dee, Flood, etc. and at present Doctor Harvey, who deserves for his many and eminent Discoveries, to have a Statue erected rather of Gold than of marble.[23]

The remarks of Hamey, Fuller, and Ashmole were all written before the Restoration, but, as Ashmole pointed out, the tide was already running against such practitioners as Fludd.[24] By the end of the century, when the names of Hobbes, Locke and Newton are looming large, and there is a well-established Royal Society to oversee and disseminate officially accepted scientific work, Robert Fludd's days of glory in England have long passed. Sir William Temple summed up his fate somewhat sympathetically in his *Essay Upon the Ancient and Modern Learning* of 1690:

> It is by themselves [the British] confessed, that, till the new philosophy had gotten ground in these parts of the world, which

is about fifty or sixty years date, there were but few that ever pretended to exceed or equal the ancients; those that did were only some physicians, as Paracelsus and his disciples, who introduced new notions in physic, and new methods of practice, in opposition to the Galenical; and this chiefly from chemical medicines or operations. But these were not able to maintain their pretence long; the credit of their cures, as well as their reasons, soon decaying with the novelty of them, which had given them vogue at first.[25]

The final and lasting blow against Fludd came in 1691: Anthony Wood (1632–1695) in that year brought out the first volume of the *Athenae Oxonienses*, which goes to 1640. There were two later editions of this much-consulted work, in 1721 and 1813–20, the latter with additions and a continuation by Philip Bliss. By now the prejudices against Fludd have come to the fore, and remained locked there for nearly three centuries, primarily due to this article:

ROBERT FLUDD, or DE FLUCTIBUS, second, afterwards eldest, son of sir Tho. Fludd, knight, sometimes treasurer of war to Q. Elizabeth in France and the Low-Countries, grandson of Dav. Fludd of Shropshire, was born at Milgate in the parish of Bearsted in Kent, became convictor (or commoner) of S John's coll. in 1591, aged 17, took the degrees in arts, studied physic, travelled into France, Spain, Italy, and Germany for almost six years. In most of which countries he became acquainted with several of the nobility of them, some of whom he taught, and for their use made the first ruder draughts of several of his pieces now extant. After his return, he, as a member of Ch. Ch., proceeded in the faculty of physic, an. 1605. About which time he practiced in London, and became a fellow of the coll. of physicians there. He was esteemed by many scholars a most noted philosopher, an eminent physician, and one strangely profound in obscure matters.

He was a zealous brother of the order of Rosa-Crucians, and did so much doat upon the wonders of chymistry, that he would refer all mysteries and miracles, even of religion, unto it, and to that end fetch the pedigree of it from God himself in his holy word. Nay he did so much prophane and abuse the word by his ridiculous and senseless applications and interpretations, in which none hath exceeded more (even to the height of

blasphemy) than he, that the learned Gassendus could not
otherwise but chastise him for it, as others since have done. His
books which are mostly in Latin are many and mystical: and as
he wrote by clouding his high matter with dark language, which
is accounted by some no better than canting, or the phrase of a
mountebank; so he spoke to his patients, amusing them with I
know not what, till by his elevated expressions he operated into
them a faith-natural, which consequently contributed to the well
working of physic.

They are looked upon as slight things among the English,
notwithstanding by some valued, particularly by Seldon, who
had the author of them in high esteem. The foreigners prize and
behold them as rarities, not that they are more judicious than
the English, but more inquisitive in such difficulties, which hath
been the reason why more of them have been printed more than
once, the titles of which, and the rest, are as follows . . . [26]

It will be noted that Wood paraphrased and embellished some of
Fuller's statements; this is of interest because these two articles
provided the basic information about Fludd for the next two
centuries, and are still often referred to today. Subsequent
compilers of biographies seemed to feel compelled to follow Wood's
lead and further embellish Fuller's words, such as in Granger's
Biographical History of England, 1775:

> Robert Fludd, second son of Sir Thomas Fludd, a treasurer of
> war to queen Elizabeth, was a celebrated physician and
> Rosicrucian philosopher. He was an author of a peculiar cast,
> and appears to have been much the same in philosophy, that the
> mystics are in divinity; a vein of unintelligible enthusiasm runs
> through his works. He frequently used this sublime cant when he
> addressed himself to his patients, which had sometimes a good
> effect in raising their spirits, and contributed greatly to their
> cure . . . The prints in his large work, intitled, "Nexus utriusque
> Cosmi," etc. are extremely singular, and only to be understood
> by a second-sighted adept. Ob. 1637, AEt. 70. See more of him
> in the "Athenae Oxoniensus."[27]

The same was true of John Aiken, M.D., who included Fludd in
a biographical collection in 1880, where he is called a "physician
and philosopher of a peculiar stamp," who was said to have used

some "kind of sublime unintelligible cant to his patients." Aiken considered it "tedious and idle to analyse all his fancies, which however, supported by mystic gravity and a shew of learning, attracted the notice of the philosphers of the day." One of Fludd's works, Aiken says (the *Nexus utriusque Cosmi*), is "illustrated (if it may be so called) by some extremely singular prints." The reader is then referred to Wood.[28]

So it went until 1861, when it got worse. In that year, William Munk brought out his collection of biographies of members of the Royal College of Physicians, a work which was enlarged and reprinted in 1878 and became a standard reference. Munk includes in his articles on Fludd entries from the *Annals* of the College, detailing his difficulties with the examiners when applying for a license and then membership. He adds a sentence which has shown up in a number of subsequent discussions of Fludd:

> With a large share of egotism and assurance, a strong leaning to chemistry, a contempt of Galenical medicine, and let us hope a sincere belief in the doctrines of the Rosie cross, absurd as these are represented to have been, he seems to have startled the Censors by his answers within the College, no less than by his conduct out of it, and was for some time in constant warfare with the collegiate authorities, and an object of deserved suspicion to his seniors in the profession.[29]

After giving the exact quotations in Latin with all the details of Fludd's disagreements with the Censors, Munk adds briefly and without comment that "on the 25th June, 1608, he was actually admitted a Candidate, and on the 20th September, 1609, a Fellow of the College. He was a Censor in 1618, 1627, 1633, 1634".[30] The article concludes with a paraphrased combination of Aiken and Granger and the Latin text of Hamey's remarks. One of the places the above uncomplimentary sentence, and the equally unflattering biographical paraphrase, have shown up is in a recent book (1964) put out by the Royal College of Physicians itself. Apparently some benefactor had Fludd's portrait painted from an engraving about the turn of the twentieth century, and it is included in the book, *The Royal College of Physicians: Portraits*, along with a brief biographical sketch taken word for word from Munk.[31]

In 1887, Fludd's first widely-read English sympathizer in over two hundred years appeared in print. A. E. Waite, occultist, writer and professional translator of occult works, including those of Paracelsus, produced *The Real History of the Rosicrucians* in that year.[32] The late 1800s saw a great resurgence of interest in the occult, spiritualism, psychic phenomena and theosophy. Waite's book was, of course, meant for this readership, and as such no doubt sold well; but it also served another purpose. Waite had at least gone to a little trouble to look up a few facts about Fludd's family background and his monument in the Bearsted, Kent parish church, which is more than any other writer had done in two centuries. The bit of factual information included in Waite's book is still referred to,[33] and may have had some influence in causing the first book on Fludd to appear fifteen years later.

The reverend James Brown Craven, rector of an Episcopal church in Kirkwall, Orkney Islands, published a book on Fludd in 1902.[34] While this study is generally even-handed, it does contain some errors, and it uncritically accepts things others have said about Fludd. It also contains a useful summary of Fludd's works and some previously buried biographical information, but much of the analysis has been outmoded by modern scholarship. Until the 1960s, there were still few serious studies which considered Fludd's actual work without excessive bias, either modern-scientific or occultist.[35] C. H. Josten, Peter Ammann, Walter Pagel, Frances Yates and Allen Debus must be singled out as the first modern scholars, whose efforts did much to increase our understanding of Fludd's works and his proper historical perspective.[36] Frances Yates' *Giordano Bruno and the Hermetic Tradition* (1964)[37] touched off a virtual renaissance of interest in Fludd, including my own, and Allen Debus' recent articles and books have done much to restore Fludd's rightful place as an important medical and scientific figure of his day.

With the present study, and others to follow, perhaps a more balanced view of Fludd's time and its relation to our own may be established, which may then allow us to incorporate the good from that era which we lost in our headlong rush into modernity; the combination would affect our vision of the future that we create, and give it a wholeness that it surely lacks.

NOTES

CHAPTER I

1 Bishop's transcripts of the Bearsted parish church register in the Cathedral Archives and Library, Canterbury. There is also a copy in the Kent Archives Office, Maidstone.
2 *The Visitation of Shropshire*, eds. G. Grazebrook & J. P. Rylands, Publications of the Harleian Society, XXIX (London: The Harleian Society, 1889), II, p. 335.
3 *Dr. Fludd's answer unto M. Foster*, p. 5.
4 *Ibid.*, pp. 6–7.
5 Jean-Jacques Boissard, *Bibliotheca sive thesaurus virtutis et gloria*, (Francofurti: Artificiossissime in aes incisae a loan. Theodor. de Bry, Sumptimus Guilielmi Fitzeri, 1628–30), Part 2, p. 198. Boissard died in 1602; this is a continuation of his format by the main publishers of Fludd's works.
6 Sir Thomas is listed as a lawyer in the Court of Common Pleas in January 1599/1600 in APC, New Series, XXX, 1599–1600, p. 30. My thanks to Judith Maizel for this information.
7 CSPF, Elizabeth I, 1562, p. 322; monument in Bearsted parish church.
8 William Lambarde, *A Perambulation of Kent* (London: Baldwin, *et al.*, 1826), p. 23; and the monument in Bearsted church.
9 CSPF, Elizabeth I, LV, 1566–1569, p. 168. He relinquished the post on 19 November 1590; CSPD, Elizabeth I, 1590, p. 698.
10 CSPD, Elizabeth I, 1581–1590, p. 436. His account books are noted in LI, XXXV, p. 321.
11 CSPD, Elizabeth I, 1581–1590, p. 618 and APC, New Series XVIII, pp. 117–18, *passim*.
12 CSPD, Elizabeth I, 1581–1590, pp. 293 and 332.
13 Examples are found in CSPD, Elizabeth I, 1591–1594, pp. 1 and 566; *ibid.*, 1595–1597, pp. 239 and 331; and HMC, 5th Report, 1876, p. 140.
14 CSPD, Elizabeth I, 1595–1597, p. 378. For more details on Sherley, see D. W. Davies, *Elizabethans Errant: The Strange Fortunes of Sir Thomas Sherley and his Three Sons* (Ithaca: Cornell University Press, 1967).
15 This interesting story of late Elizabethan in-fighting at court is chronicled in: CSPD, Elizabeth I, 1595–1597, pp. 379 and 402; HMC Salisbury MSS, Part VII, pp. 130, *passim*; and HMC Lord de L'Isle and Dudley MSS, 3, pp. xxix and xli.
16 HMC, Lord de L'Isle and Dudley MSS, 3, p. 74.
17 "Return . . . from the year 1213 . . . to the year 1696 of the Surnames, Christian names, and titles of all Members of the Lower House of Parliament . . . " Great Britain. Parliament. House of Commons. Sessional Papers, 1878, vol. LXII, part 1; pp. 148, 433 and 438. I thank Judith Maizel for this lead.

NOTES

18 *The Register of Admissions to Gray's Inn, 1521–1889,* Joseph Foster, ed. (London: Privately Printed, 1889), p. 101.

19 *The Visitation of Kent, 1530–1 and 1574,* W. B. Bannerman, ed., Harleian Society Publications LXXIV (London: The Harleian Society, 1923), p. 50. *The Visitation of Kent, 1592,* part 2, W. B. Bannerman, ed., Harleian Society Publications LXXV (London: The Harleian Society, 1924), p. 99.

20 Probate copy of Sir Thomas' will, Public Record Office, London: Prob 11/110 LH 55-LH 59.

21 Elizabeth Fludd died 25 January 1592 (monument in Bearsted Church).

22 See J. H. Cooper, "The Coverts, Part II," *Sussex Archaeological Collections,* XLVII (1904), pp. 139–40; and *The Visitations of Essex,* Harleian Society Publications XIII (London: The Harleian Society, 1878), p. 28. Lady Barbara died in 1619; *Index of Wills Proved in the Prerogative Court of Canterbury,* The Index Library V (London: The British Record Society, 1912), p. 169.

23 Probate copy of Robert Fludd's will, Public Record Office, London, Prob. 11/175 LH 18-RH 19.

24 The nearby (16 miles) school of Tonbridge elected one Fellow to St John's College, Oxford, but there is no present evidence that Fludd attended school there.

25 *Register of the University of Oxford,* ed. Andrew Clark, vol. II, part II, "Matriculations," Oxford Historical Society Publications, XI (Oxford: Oxford Historical Society, 1887), p. 193.

26 Mark Curtis, *Oxford and Cambridge in Transition 1558–1642* (Oxford: The Clarendon Press, 1959), p. 54; Joan Simon, *Education and Society in Tudor England* (Cambridge: The University Press, 1966), pp. 59ff.

27 Curtis, p. 17; Simon, pp. 40–2.

28 Training in the law at the Inns of Court was the major alternative for advanced education. Simon, pp. 16–17, 43, 353ff. See also Lawrence Stone, "The Size and Composition of the Oxford Student Body 1580–1609," *The University in Society,* ed. Lawrence Stone, 2 vols. (Princeton: Princeton University Press, 1974), I, pp. 18–19; and Curtis, p. 58.

29 Simon, pp. 59ff.; Curtis, pp. 70–1.

30 *Ibid.*

31 *Ibid.*

32 Curtis, p. 72; Simon, pp. 353ff.; Stone, pp. 24–8.

33 Stone, pp. 20 and 93; Curtis, p. 59.

34 Curtis, p. 31.

35 *Ibid.,* p. 37. In the case of St John's, M.A.s were to take orders within three years of the beginning of theological studies, with the Medical Fellow and Public Readers excepted. See below, Mallet, p. 179 and Stephenson and Salter, p. 145.

36 W. H. Stephenson and H. E. Salter, *The Early History of St. John's College, Oxford,* OHS New Series, vol. I (Oxford: at the Clarendon Press, 1939), pp. 113ff. W. C. Costin, *The History of St. John's College,*

Oxford 1598–1860, OHS New Series, vol. 12 (Oxford: at the Clarendon Press, 1958), pp. 1ff. C. H. Mallet, *A History of the University of Oxford* (New York: Longmans, Green, 1924), II, pp. 174ff.

37 Costin, p. 5.

38 *Ibid.*, p. 4.

39 *Ibid.*, p. 8.

40 *Ibid.*, p. 2.

41 Stephenson and Salter, pp. 301–19; Costin, p. 6.

42 Mallet, p. 177 n. 4.

43 *Ibid.*, p. 123. In 1589 the Convocation decreed that no student could advance to the Bachelor's or Master's degree unless he could recite the Articles of Faith from memory. *Ibid.*, p. 122.

44 Curtis, p. 80.

45 John Perrin was born in London, and entered Merchant Taylors' School on 17 October 1572. From there he was elected to a Fellowship at St John's, Oxford, matriculating in 1575 at the age of seventeen. He received his B.A. in 1580, M.A. in 1583, and B.D. in 1589. In 1588 Perrin was appointed to the Greek lectureship of the College, which he held until 1605, when he resigned to take the post of the King's Professorship of Greek for the University. The doctorate of divinity was conferred upon him in 1596. He published five poems, all in Oxford collections: the first in Latin on the death of the Lord Chancellor Hatton in 1592; the second and third to the memory of Queen Elizabeth, one in Greek and one in Hebrew, in 1603; the fourth in Greek on James I's accession; and the fifth, also in Greek, in honor of a royal visit to Oxford in 1605. Perrin held the vicarage of Watling, Sussex (1605–1611), and was a canon of Christ Church cathedral, Oxford. He died on 9 May 1615, and his will was proved in the Vice-Chancellor's court on 25 May 1616. On his assumption of the King's public lectureship in Greek at the University, he was also employed in the translation of the New Testament.

   Fludd mentioned him as his tutor in *Utriusque cosmi . . . historia*, I, 2, "De Naturae Simia," p. 701. See also Anthony Wodd, *Fasti Oxonienses*, I, 273; *Register of the University of Oxford*, i, 198, ii, 69, iii, 83; *Register of the Merchant Taylors' School*, p. 20; Stephenson & Salter, pp. 350–1; Falconer Madan, *Oxford Books*, vol. 2, pp. 22, *passim*; Frances Yates, *Theater of the World* (Chicago: The University of Chicago Press, 1969), p. 61; and Peter J. Ammann, "The Musical Theory and Philosophy of Robert Fludd," JWCI, 30 (1967), 198–227.

46 Costin, p. 16.

47 *Ibid.*, p. 17.

48 *Ibid.*, p. 57.

49 *Ibid.*, pp. 57–8.

50 Curtis, pp. 151–2.

51 From the data given in William Munk, *Roll of the Royal College of Physicians in London* (London: Longman, *et al.*, 1861), vol. I.

52 *Ibid.*

53 *Ibid.*, p. 114. See also John Ward, *The Lives of the Professors of Gresham*

College (New York: Johnson Reprint, 1967), pp. 260–5.

54 Allen Debus, *The English Paracelsians*, pp. 142–3.

55 Stephenson and Salter, p. 343.

56 *Ibid.*; DNB X, p. 343.

57 DNB VIII, pp. 242–3.

58 Stephenson and Salter, pp. 369–70.

59 Fludd, *The Philosophical Key*, op. cit., fol 15ᵛ.

60 This is from John Case, *Speculum Moralium Quaestionum*, Oxford, 1585; translated in W. S. Howell, *Logic and Rhetoric in England, 1500–1700* (Princeton: The University Press, 1956), pp. 190–1.

61 *Ibid.*, p. 178.

62 W. C. Costin, "The Inventory of John English, B.C.L., Fellow of St. John's College," *Oxoniensia*, XI & XII (1946–7), 102–31.

63 Fludd, *Mosaicall Philosophy*, p. 36.

64 Fludd, "De Naturae Simia," p. 701.

65 That Fludd was exposed to Neoplatonist-Hermetic interests at Oxford is certain; see Mordechai Feingold, "The occult tradition in the English universities of the Renaissance: a reassessment," in Brian Vickers, ed., *Occult and Scientific Mentalities in the Renaissance* (Cambridge: Cambridge University Press, 1984).

66 *Register of the University of Oxford*, II, 3, p. 195.

67 Costin, *History of St. John's*, p. 58.

68 Fludd, "De Naturae Simia," p. 3. Exactly when Fludd travelled in Spain is not clear, since England and Spain were at war from 1588 to 1604.

69 In Paris: Fludd, *Anatomiae Amphitheatrum*, p. 233. English translation in J. Webster, *A Displaying of Supposed Witchcraft* (London: Jonas Moore, 1677), pp 319–20. In Lyons: *Mosaicall Philosophy*, p. 277.

70 Fludd, "De Naturae Simia," pp. 717–20. C. H. Josten has translated this entire passage into English in "Robert Fludd's Theory of Geomancy and his Experiences at Avignon in the Winter of 1601 to 1602," JWCI, XXVII (1964), 372–5. See also Frances Yates, *Theater of the World*, op. cit., p. 63. Fludd also says he taught musical theory to the Marquis de Orizon, Viscount de Cadenet. See Ammann, op. cit., p. 205.

71 Fludd, *Mosaicall Philosophy*, pp. 236–7.

72 Fludd, *Dr. Fludds answer unto M. Foster*, p. 134. Gruter is discussed further in Chapter III, note 35.

73 *Mosaicall Philosophy*, p. 100.

74 Frances Yates' *Rosicrucian Enlightenment* (London: Routledge & Kegan Paul, 1972) considers some intriguing aspects of the Palatinate court as an ideological center of Protestant Continental politics.

75 *Ibid.*, pp. 70ff.

CHAPTER II

1 W. C. Costin, *The History of St. John's College, Oxford 1598–1860* (Oxford: Clarendon Press, 1958), p. 25; and J. F. D. Shrewsbury, *A History of*

Bubonic Plague in the British Isles (Cambridge: The University Press, 1970), p. 282.

2 Clark, *Reg. Univ. Oxford*, II, 3, p. 194.

3 Curtis, pp. 152–3.

4 Clark, *Reg. Univ. Oxford*, II, 1, p. 191:
   I. In Vesperiis Roberti Fludd
      1. "Frequens usus purgantium
         medicamentorum non accelerat
         sensium."
      2. "Chymicum extractum minus
         ımolestiae et periculi affert
         quam quod integrum et naturale."
      3. "Senes facilius inediam ferunt
         quam pueri."

5 Keynes, *The Life of William Harvey* (Oxford: Clarendon Press, 1966), pp. 134–5.

6 *Ibid.*

7 Curtis, pp. 153–4.

8 Debus, *The English Paracelsians* (New York: Franklin Watts, 1972), p. 29 *passim.*

9 Munk, p. 141:
"Secundo examinatur, atque etiamsi plene examinationibus non satisfaceret, tamen judicio omnium visus est non indoctus, permissus est itaque illi medicinam facere."

10 Munk, pp. 141–2:
"Delatum est ad Collegium D^m Fludd multa de se et medicamentis suis chemicis praedicasse, medicos autem Galenicos cum contemptu dejecisse; Censores itaque in hunc diem citari cum jesserunt. Interrogatus ad id verum esset, quod objectum est, confidentissime omnia negabat, et accusatores requirebat; qui quoniam non comparebant dismissus est cum admonitione, ut modeste de se et sentiret et loqueretur; Socios autem Collegii reveraetur. Et cum persolvisset pensionem a statutis praescriptam, admissus est in numerum Permissorum."

11 Munk, p. 142:
"22 December 1607. Dr. Fludd, examinatus, censetur dignus qui fiat Candidatus."

12 Munk, p. 142:
"Dr. Fludd, qui jam in Candidatorum numerum erat cooptandus, tam insolenter se gessit, ut omnes offenderantur, rejectus est itaque a D^o Praesidente cum admonitione, ut sibi, si sine Licentia practicare pergeret, diligenter caveret." According to Keynes, William Harvey was present at this meeting; Keynes, p. 135.

13 Munk, p. 142.

14 Sir George Clark, *A History of the Royal College of Physicians of London* (Oxford: Clarendon Press, 1964), II, p. 223.

15 *Annals of the College of Physicians*, ii, 27 June 1620. I am indebted to Judith Maizel for this information.

16 APC, 1627, January-August, pp. 444–5.

17 On 25 June 1635. Sir George Clark, p. 253.

18 *Ibid.*, p. 185.

19 College of Physicians, *Pharmacopoeia Londinensis* (London: E. Griffin, 1618).

20 W. C. Underwood, *A History of the Worshipful Society of Apothecaries of London* (London: Wellcome Museum, 1963), I, pp. 49, *passim.*

21 *A Physical Directory, or a translation of the London Dispensatory made by the College of Physicians in London* (London: Peter Cole, 1649).

22 *Mr. Culpepper's Ghost* (London: for Peter Cole, 1656).

23 *Ibid.*, pp. 1–3.

24 Theodore Goulston (1572–1632), M.A. 1600, M.D. 1610 from Merton College, Oxford, Fellow of the College of Physicians 1611–1632, and "an excellent Latinist, noted Grecian, but better for theology," Munk, p. 147. Fludd owned Goulston's translation of Aristotle's *Rhetoric*, published in 1619 in London, which he gave to Jesus College, Oxford in 1630; as recorded in The Benefactors' Book.

25 *Mr. Culpepper's Ghost*, pp. 5–9. A longer portion of this passage includes a fictitious and amusing debate between Fludd, Lull and Van Helmont about the weapon-salve. Since they could not agree, arbitrators were chosen: Fludd picked Trismegistus and Van Helmont chose Paracelsus, but the matter remained unresolved. See F. N. L. Poynter, "Nicholas Culpepper and the Paracelsians," *Science, Medicine and Society in the Renaissance*, ed. Allen Debus (New York: Science History Publications, 1972), pp. 217–19.

26 Allen Debus, *The English Paracelsians*, p. 47.

27 *Ibid.*, p. 51.

28 For an elaboration of Fludd's medical theories, see Walter Pagel, "Religious Motives in the Medical Biology of the XVIIth Century," *Bulletin of the Institute of the History of Medicine*, 3 (1935), 265–92; see also Fludd, *The Philosophical Key*, fols 54^r–57^v; *Mosaicall Philosophy*, pp. 82–96; *Medicina Catholica, passim; Philosophia Sacra, passim.* For an excellent series of illustrations with explanations of the winds and related schemes, see Joscelyn Godwin, *Robert Fludd: Hermetic Philosopher and Surveyor of Two Worlds* (London: Thames & Hudson, 1979; Boulder: Shambhala, 1979), pp. 54–67.

29 Fludd had illustrations of himself, elegantly attired in wide-brim hat and cape, inserted in his *Integrum Morborum Mysterium* and *ΚΑΘΟΛΙΚΟΝ*, showing him taking a pulse at the bedside, casting a horoscope and examining a urine sample. See Plates 14, 15, 16.

30 The central argument about the "weapon-salve" and other such methods is whether cures can be effected from a distance through a sympathetic harmony, and if so, whether these cures are diabolical in origin. Followers of Aristotle believed that there can be no action remote from that which causes the action. When Fludd's weapon-salve was attacked by William Foster in 1631 (see Chapter I), it was *not* on the grounds that it did not work, but that it was diabolical. A good summary of the weapon-salve controversy is in Allen Debus, *The*

Chemical Philosophy: Paracelsian Science and Medicine in the Sixteenth and Seventeenth Centuries (New York: Science History Publications, 1977), pp. 246–50.

31 Barber-Surgeons, Court Minutes, 1621–1651, pp. 123 and 171. See R. S. Roberts, "The Personnel and Practice of Medicine in Tudor and Stuart England. Part II. London," *Medical History* 8 (1964), 233–4, n. 93. I am indebted to Judith Maizel for this reference. Andrewes was sworn in on 25 June 1631 and Fludd on 12 June 1634. Dates kindly supplied by Godfrey Thompson, Guildhall Library.

32 Fludd, *Clavis Philosophiate et Alchemiae Fluddanae* . . . (Frankfurt: Wilhelm Fitzer, 1633), pp. 33ff. See also Allen Debus, *The Chemical Philosophy*, p. 286.

33 Keynes, p. 100.

34 Fludd, *Anatomiae Amphitheatrum* (Frankfurt: de Bry, 1625), p. 101.

35 An interesting article in this regard is H. P. Bayon, "William Gilbert, Robert Fludd and William Harvey as Medical Exponents of Baconian Doctrines," *Proceedings of the Royal Society of Medicine, XXXII* (1938–1939), 31–42.

36 Fludd, "The Philosophical Key," fol. 56^v.

37 APC, 1617–1619, p. 135. My thanks to Judith Maizel for calling my attention to this reference.

38 *Ibid.*, p. 212.

39 *Ibid.*, pp. 284–5. For details of Fludd and steel-making in this era, see J. W. Gough, *The Rise of the Entrepreneur* (New York: Schocken Books, 1969), Chapter 4, "Iron and Steel." What became of his plans to make steel is not presently known.

40 Daniel Sennert, *The Weapon-Salves Maladie* (London: John Clark, 1637), sig. A3^v. Quoted in Debus, *The Chemical Philosophy*, Chapter 4.

CHAPTER III

1 *Register of the University of Oxford*, II, p. 185. Stephenson & Salter say he came from Merchant Taylors' School, but he is not listed in the printed register; there is a "Christopher Andrew" registered for 25 February 1585/1586, but it seems doubtful that this is the same person.

2 *Ibid.*, iii, p. 192.

3 W. H. Stephenson and H. E. Salter, *The Early History of St. John's College, Oxford* (Oxford: Clarendon Press, 1939), p. 369.

4 *Register of the University of Oxford*, III, p. 192.

5 Munk, p. 145.

6 *Ibid.* The appointment was to take effect on Harvey's death, but Andrewes died in 1634, twenty-three years before Harvey.

7 Fludd, "The Philosophical Key," fol. 15^v. See transcription with introduction by Allen Debus, *Robert Fludd and His Philosophical Key* (New York: Science History Publications, 1979), p. 73.

8 *Register of the University of Oxford*, III, p. 183.

9 DNB II, p. 704. It was during this service that Laud married
Mountjoy and his divorced mistress, an act that haunted him for the
rest of his life and kept him out of favor with James I. See
H. R. Trevor-Roper, *Archbishop Laud* (London: Macmillan, 1940),
pp. 11–13.

10 Trevor-Roper, p. 11; W. C. Costin, *The History of St. John's College,
Oxford* (Oxford: Clarendon Press, 1958), pp. 26–34; and W. H. Hutton,
St. John Baptist College, Oxford University College Histories (London:
Robinson, 1898), pp. 86–121.

11 Fludd's *Integrum Morborum Mysterium* (Frankfurt: William Fitzer, 1631)
is dedicated to George Abbot: "Reverentiae tuae servus & cultor
humillimus – R. F." (This is the second treatise of his *Medicina
Catholica*). The dedication to Williams is at the beginning of the
Philosophia Sacra . . . of 1626, probably written in 1625 before Williams
was removed as Lord Keeper (on 25 October) after James I's death.

12 DNB I, p. 7.

13 W. H. Hutton, p. 99.

14 DNB I, p. 7.

15 *Ibid.*, p. 6.

16 Trevor-Roper, pp. 52–62; DBN XXI, pp. 441ff.

17 Fludd, "Declaratio Brevis," British Museum MS Royal 12.c.ii, fol.1^v.
The "Declaratio" was written about 1618 and addressed to James I,
but it is not known whether the King actually read it. See Chapter IV
and Appendix A.

18 Fludd, "The Philosophical Key," fol. 7^v; Debus, p. 69.

19 Fludd, "Declaratio Brevis," fol. 1^v.

20 HCM9, Salisbury MSS XVIII, pp. 241–2.

21 Fludd, *Anatomiae Amphitheatrum*, p. 233. English translation in John
Webster, *A Displaying of Supposed Witchcraft* (London: Jonas Moore,
1677), pp. 319–20.

22 Fludd, "Declaratio Brevis," fol. 6^v.

23 *Ibid.* The letter is dated 19 December 1617.

24 Fludd, *Utriusque cosmi . . . historia*, I, 2, "De Naturae Simia," p. 720; and
C. H. Josten, "Robert Fludd's Theory of Geomancy and his
experiences at Avignon in the winter of 1601 to 1602," JWCI, 27
(1964), p. 335.

25 H. Forneron, *Les Ducs de Guise et leur époque*, II (Paris: E. Plon, 1877),
pp. 398–9.

26 Philip wanted Isabella to marry the Holy Roman Emperor, but the
Estates-General refused to set aside the Salic Law to permit the
accession. Forneron, pp. 405–6.

27 Charles of Lorraine, fourth Duke of Guise, commanded an army from
Champagne against Henry of Navarre after his escape, but submitted
to the new King in 1594 (22 October). In return he was made
Governor of Provence, Admiral of the Levant, a "grand-maitre de
France," and received a pension of 400,000 crowns to support four
companies. In addition, his brothers received the Archbishopric of
Reims, five large abbeys, plus pensions and allowances. During the

regency following Henry IV's assassination in 1610, Charles aligned himself with the Queen Mother's faction. After Richelieu's rise to power in 1624, inevitable distrust and power struggles ensued, with relations between Guise and the first minister becoming very bitter. After a power showdown in 1631 with Richelieu victorious, the Duke went into voluntary exile in Italy, where he died in 1640. Forneron, pp. 417, 428–30. (Charles and his brother also happened to be third cousins of James VI of Scotland).

28 François was accidentally killed on 1 June 1613 when he personally put the torch to a cannon he had overloaded, causing it to blow apart. Forneron, p. 426. Fludd laments his loss with a strong expression of sorrow ("proh dolor!") in "De Naturae Simia," p. 3. He was only twenty-four at the time of his death, and a youth of twelve when Fludd's pupil.

29 *Utriusque cosmi . . . historia*, I, 2, "De Naturae Simia," p. 3.

30 *Ibid.*

31 See Peter J. Ammann, "The Musical Theory and Philosophy of Robert Fludd," JWCI, 30 (1967), pp. 198–227. I have been unable to further identify Viscount Cadenet.

32 For details of Fludd's geomancy and his conversations with the Vice-Legate, see Fludd, "De Naturae Simia," pp. 717–20; and C. H. Josten, article cited, pp. 332–5.

33 Fludd, "De Naturae Simia," pp. 3 & 718.

34 *Ibid.*, p. 718; Josten, p. 33.

35 Fludd, *Answer to M. Foster*, pp. 134–5. The "Master Gruter" referred to is unknown. A possible but unlikely identification is that of Janus Gruterus (1560–1627), the famous humanist philologist and head of the Palatinate library in Heidelberg after 1602. It is quite possible that Fludd met Gruterus in Heidelberg – Gruterus' mother was English, he attended Cambridge in 1577, and carried on a wide correspondence, which included William Camden. However, Fludd's Master Gruter was born in Switzerland, and Gruterus in Antwerp, so the two do not appear to coincide. See *Allgemeine Deutsche Biographie*, 10, pp. 68–71; and *Neue Deutsche Biographia*, 7, pp. 238–40; and *Biographie Nationale*, 7, cols 363–81.

36 Frederick V (1596–1632), the Elector Palatine following his father's death in 1610, married James I's daughter Elizabeth on 14 February 1613. Both Frederick IV and V were Calvinists and continually maintained close ties with Elizabeth I and James I. See Frances Yates, *The Rosicrucian Enlightenment* (London and Boston: Routledge & Kegan Paul, 1972).

37 See Christopher von Rommel, *Geschichte von Hessen*, 9 vols. (Marburg and Cassel: Fr. Perthes *et al.*, 1820–1853), VI, pp. 382–782; Karl E. Demandt, *Geschichte des Landes Hessen* (Kassel & Basel: Barenreiter Verlag, 1972), pp. 246–62.

38 Landgrave William endowed a chair of mathematics at the University of Marburg and made the subject a required part of the curriculum for philosophical studies. See von Rommel, V, pp. 758–92; and Herman

Schlenz, "Goldmachen und Goldmacher am hessischen Hofe," *Deutsche Geschichtsblatter*, 11 (1910), 308–11.

39 Schlenz, pp. 309 and 310.

40 Landgrave Moritz's son Otto paid a two-month visit to England in 1611, when he was warmly received and entertained by James I and Prince Henry. Philipp Losch, "Die Reise des Landgrafen Otto von Hessen nach England und den Niederlanden in Jahre 1611," *Hessenland*, 10 (1931), 289–96.

41 The only monograph on Maier is J. B. Craven, *Count Michael Maier* (Kirkwall, Orkney Islands: Wm Peace, 1910). Maier and Fludd shared philosophy and publishers: see Frances Yates, *Theater of the World* (Chicago: University of Chicago Press, 1969), pp. 65 and 72; and by the same author, *The Rosicrucian Enlightenment* cited above, pp. 70–91, *passim*. On Maier taking Fludd's *Macrocosm* to de Bry, see Chapter VIII.

42 Craven, p. 8; see Chapter VIII.

43 At Oxford Fludd worked on his cosmography and a treatise on music. Fludd, "De Naturae Simia," p. 3.

44 Dedication by "R. N. E." in John Baptista Lambye, *A Revelation of the Secret Spirit declaring the most concealed secret of Alchemia*, 1623, trans. R. N. E. Quoted in Allen Debus, *The English Paracelsians*, pp. 104–5.

45 Fludd, *Anatomiae Amphitheatrum*, 1623: "Venerando atque reverendo viro, et in accurata Naturae mysteriorum inquisitione studiosissimo, Ioanni Thornburgh, Episcopo Wigorniensi amico meo singulari, in eo ipso qui est vera mundi lux, et thesaurorum thesaurus, salutem."

46 John Thornborough's references to Fludd are in his λιθοθεωρικοσ, sive, Nihil, Aliquod, Omnia, Antiquorum Sapientum Viviscoloribus depicta (Oxoniae: Johannes Lichfield and Jacobus Short, 1621), pp. 60, 68, 126 and 127. Compare, for example, with Fludd, *Tractatus Theologo-Philosophicus* (Oppenheim: de Bry, 1617), pp. 72, 98, *et passim*.

47 Fludd, *Mosaicall Philosophy*, p. 118. The Mr Finch referred to is not further identified, but he could have been a member of the prominent Kentish family of the same name.

48 See DNB VII, "Forman, Simon," p. 438. Forman was in Thornborough's service 1573–1574.

49 DBN IV, p. 1233.

50 Joan Evans, *A History of the Society of Antiquaries* (Oxford: The University Press, 1956), pp. 10 and 13.

51 DNB IV, p. 1234; DNB XVII, p. 1150.

52 DNB IV, p. 1234.

53 John Dee, *A True and Faithful relation of what passed for many yeers between Dr. John Dee and some spirits*, preface by Meric Casaubon (London: Maxwell for T. Garthwait, 1659), pp. 44 of the preface.

54 DNB XVII, p.1151.

55 John Selden, *Titles of Honor* (London, 1614), A3.

56 Fludd, ΚΑΘΟΛΙΚΟΝ (Frankfurt: W. Fitzer, 1631).

57 See Chapter III.

NOTES

58 See Walter Pagel, *William Harvey's Biological Ideas* (Basel & New York: S. Karger, 1975).
59 DNB xvii, p. 1152.
60 See Chapter IV.
61 *Ibid.*
62 Fludd, "The Philosophical Key," fol. 53^v; Debus, p. 97.
63 Fludd, *Dr. Fludds answer unto M. Foster*, p. 21.
64 CSPD, Charles I, 1628–1629, p. 570.

CHAPTER IV

1 Fludd, "Declaratio Brevis," British Library, Royal MS. 12 C ii. Translated by Robert Seelinger and William H. Huffman, with an introduction by William H. Huffman, *Ambix* 25 (1978), 69–92.
2 See Chapter VIII.
3 See Bibliography.
4 Fludd, "A Philosophical Key," Trinity College, Cambridge, Library, Western MS. 1150. Published edition: Allen G. Debus, *Robert Fludd and His Philosophical Key* (New York: Science History Publications, 1979), fol. 15^v.
5 *Ibid.*, fol. 14^v.
6 Fludd, *Doctor Fludds Answer unto M. Foster* (London: Nathanael Butter, 1631), p. 22.
7 Fludd, "Declaratio Brevis," Fol. 1^r.
8 *Ibid.*
9 *Ibid.*, fols 1^r–1^v.
10 *Ibid.*, fol. 2^r.
11 *Ibid.*, fols. 1^v–2^r.
12 *Ibid.*, fols. 2^v–3^r.
13 *Ibid.*, fols. 3^r–3^v.
14 *Ibid.*, fol. 4^v.
15 *Ibid.*, fols. 5^r–5^v.
16 *Ibid.*, fol. 5^v.
17 *Ibid.*, fols. 6^r–9^v; on Horst, see note 10 on the *Declaratio*.
18 "Declaratio," fol. 9^v.
19 "Philosophical Key," cited above.
20 *Ibid.*, fol. 4^r.
21 *Ibid.*, fol. 6^v.
22 *Ibid.*, fols 7^v–8^r.
23 *Ibid.*, fols. 8^v–9^v.
24 *Ibid.*, fol. 15^r.
25 *Ibid.*, fol. 17^r.
26 *Ibid.*, fol. 18^v.
27 *Ibid.*, fol. 21^v.
28 See Fludd's will, Appendix B, and note 8 to the will.
29 APC, January 1618–June 1619 (London: HMSO, 1929), p. 135.
30 *Ibid.*, July 1619–June 1921 (London: HMSO, 1930), p. 212.
31 *Ibid.*, p. 284.

NOTES

32 *Ibid.*, p. 285.
33 *Ibid.*, p. 319.
34 CSPD, Charles I, 1628–1629 (London: Longman, *et al.* 1859), p. 570.
35 State Papers 39/27.29. I am grateful to Judith Maizel for this reference.

CHAPTER V

1 For an excellent selection of Fludd's illustrations with explanations, see
Joscelyn Godwin, *Robert Fludd: Hermetic Philosopher and Surveyor of Two
Worlds* (London: Thames and Hudson, and Boulder: Shambhala,
1979); a summary of Fludd's books and bibliographic data is in
J. B. Craven, *Doctor Robert Fludd* (Kirkwall: William Peace & Sons,
1902). On Fludd's publishers, see E. Weil, "William Fitzer, the
publisher of Harvey's *De motu cordis*, 1628," *The Library*, Fourth Series
24 (1944), 142–64; Moritz Sondheim, "Die de Bry, Matthäus Merian
und Wilhelm Fitzer," *Philobiblon* 6 (1933), 9–34; and
J. A. Van Dorsten, *Thomas Basson, 1555–1613: English Printer at Leiden*
(Leiden: University of Leiden, 1961).
2 Kepler to Galileo, from Graz, 13 October 1597, in Carola Baumgardt,
Johannes Kepler: Life and Letters (London: Gollancz, 1952), p. 41.
3 *Ibid.*, p. 76.
4 Arthur Koestler, *The Watershed* (New York: Doubleday Anchor, 1960),
p. 76.
5 Johannes Hemleben, *Johannes Kepler* (Reinbek bei Hamburg: Rowohlt,
1971), p. 89.
6 In addition to the three authors discussed here, see also D. P. Walker,
"Kepler's Celestial Music," JWCI 30 (1967), pp. 208–50; Frances
Yates, *Giordana Bruno and the Hermetic Tradition* (London: Routledge &
Kegan Paul, 1964), pp. 440–4; Allen G. Debus, *The Chemical Philosophy*,
2 vols. (New York: Science History Publications, 1977); Todd Barton,
"Robert Fludd's Temple of Music," M.A. Thesis, University of
Oregon, 1978; Robert Westman, "Magical Reform and Astronomical
Reform: The Yates Thesis Reconsidered," in *Hermeticism and the Scientific
Revolution* (Los Angeles: Clark Library, University of California, Los
Angeles, 1977); Brian Vickers, "Analogy versus identity: the rejection
of occult symbolism, 1580–1680," in Brian Vickers, ed., *Occult and
Scientific Mentalities in the Renaissance* (Cambridge: Cambridge University
Press, 1984). Another discussion of Fludd and Kepler may be found in
J. V. Field, *Kepler's Geometrical Cosmology* (Chicago: University of
Chicago Press, 1988), pp. 179–87.
7 Peter J. Ammann, "The Musical Theory and Philosophy of Robert
Fludd," JWCI 30 (1967), 198–227.
8 *Ibid.*, p. 213.
9 *Ibid.*
10 C. G. Jung and W. Pauli, *The Interpretation of Nature and the Psyche* (New
York: Bollingen Foundation, 1955), pp. 147–240.
11 *Ibid.*, p. 153.
12 *Ibid.*, p. 171.

13 *Ibid.*, p. 191.
14 *Ibid.*, pp. 197–8.
15 *Ibid.*, pp. 206–7.
16 *Ibid.*, p. 208.
17 Robert S. Westman, "Nature, art and psyche: Jung, Pauli, and the Kepler-Fludd polemic," in Brian Vickers, ed. *Occult and Scientific . . . , op. cit.*, pp. 177–229.
18 *Ibid.*, p. 181.
19 *Ibid.*, pp. 187–8.
20 *Ibid.*, pp. 188–9.
21 Ammann, *op. cit.*, pp. 201ff.; Westman, "Nature," pp. 191, 198–9; Fludd, *Mosaicall Philosophy* (London; Humphrey Moseley, 1659), pp. 73 *et passim.*
22 Westman, "Nature," pp. 194–200.
23 *Ibid.*, p. 206.
24 *Ibid.*, p. 212.
25 *Ibid.*
26 *Ibid.*, pp. 217–8.
27 C. G. Jung, *Psychology and Alchemy* [Collected Works, Vol. 12, Bollingen Series XX] (Princeton: Princeton University Press, 1968), p. 47.
28 *Ibid.*, p. 188; Westman, "Nature," p. 218.
29 Jung, *Psychology*, p. 188.
30 *Ibid.*
31 Hemleben, *op. cit.*, p. 109.
32 See "Declaratio Brevis," Appendix A.
33 DNB, Vol 21, pp. 966–70; Koestler, *op. cit.*, p 242; Baumgardt, *op. cit.*, pp. 147–8.
34 Patrick Scot, *The Tillage of Light* (London: William Lee, 1623).
35 C. H. Josten, "Truth's Golden Harrow: An Unpublished Alchemical Treatise of Robert Fludd," *Ambix* 3 (1949), 91–150; Debus, *op. cit.*, p. 255.
36 Lynn Thorndike, *A History of Magic and Experimental Science.* 8 vols. (New York: Columbia University Press, 1923–58), VII, pp. 426–64; Debus, *op. cit.*, pp. 260–79; Frances Yates, *op. cit.*, pp. 438–40; Robert Lenoble, *Mersenne ou la Naissance du Méchanisme* (Paris: J. Vrin, 1943); Luca Cafiero, "Robert Fludd e la Polemica con Gassendi," *Revista Critica di Storia Filosofia* 20 (1965), 4–15.
37 Quoted in Debus, p. 267.
38 *Ibid.*, p. 268.
39 Joachim Frizius, *Summum Bonum: Quod est verum* [(*Magiae, Cabalae, Alchemiae, verae*) *Fratrum Roseae Crucis verorum*] *subjectum. In dictarum Scientiarum laudem & insignis calumniatoris Fratris Marini Mersenni dedecus publicatum . . .* (n.p. [Frankfurt]: Officina Bryana, 1629).
40 Fludd, *Clavis Philosophiae et Alchemiae Fluddanae* (Frankfurt: William Fitzer, 1633), pp. 20 and 26; Frizius, *op. cit.*, p. 2.
41 Pierre Gassendi, *Epistolica exercitatio in qua principa philosophiae Roberti Fluddi, medici, reteguntur, et ad recentes illius libros R. P. F. Mersennun . . . respondetur* (Paris, 1630); Thorndike, pp. 441ff.; Debus,

pp. 269–77.
42 Debus, pp. 273–6.
43 Fludd, *Pulsus* (n.p. [Frankfurt?]: n. d [1629]) {M.C. 1,2,ii,3,3}, p. 11; quoted in Debus, pp. 274–5.
44 Yates, pp. 398–431.
45 *Ibid.*, p. 439.
46 *Ibid.*, p. 398.
47 Casaubon (1559–1614) was a Protestant Greek scholar from Geneva who lived in England from 1610–1614. James I had asked him to reply to a Counter-Reformation work on Church history, *Annales Ecclesiastici* by Cesare Baronius (12 vols., 1588–1607), but he had only completed an analysis of the first half of the first volume at his death. Embedded in the analysis is the dating of the *Hermetica*. See Yates, pp. 398–403.
48 Fludd, *Clavis*; quoted in J. B. Craven, *op. cit.*, p. 149.
49 P. Jean Durelle, *Effigies contracta Roberti Flud . . .* (Paris: Apud Guillelmum Baudry, 1636); see Debus, pp. 278–9.
50 William Foster, *Hoplocrisma-spongus: or, A sponge to wipe away the weapon-salve* (London: Thomas Cotes, 1631); Debus, pp. 279–90; Craven, pp. 198–214.
51 Fludd, *Doctor Fludds Answer unto M. Foster, or the Squeesing of Parson Foster's sponge* (London: Nathaniel Butter, 1631).
52 *Ibid.*, pp. 15–16.
53 *Ibid.*, p. 18.
54 *Ibid.*, pp. 18–19.
55 *Ibid.*, p. 21.
56 *Ibid.*, pp. 21–2.
57 *Ibid.*, p. 23.
58 *Ibid.*, p. 24.
59 *Ibid.*
60 *Ibid.*
61 See Debus, pp. 271ff.
62 Fludd, *Doctor Fludds Answer*, p. 22.
63 Fludd, *Mosaicall Philosophy*, pp. i–ii.
64 *Ibid.*, p. 300.
65 *Ibid.*

CHAPTER VI

1 Charles B. Schmitt, *Aristotle and the Renaissance* (Cambridge, Mass. and London: Harvard University Press, 1983), pp. 10–33.
2 Alan Bloom, trans., *The Republic of Plato* (New York: Basic Books, 1968), p. xviii.
3 See Paul O. Kristeller, *Renaissance Thought and Its Sources* (New York: Columbia University Press, 1979).
4 Gershom Scholem, *On the Kabbalah and its Symbolism*, tr. Ralph Mannheim (New York: Schocken Books, 1969), p. 52.

5 James M. Robinson, ed., *The Nag Hammadi Library* (New York: Harper and Row, 1981).
6 St Augustine, *The City of God* (Garden City, N.Y.: Image Books, 1958), p. 570 [Book VIII, Chapter 5].
7 St Augustine, *The Confessions of St Augustine* (New York: New American Library, 1963), pp. 156–7 [Book VII, Chapter 20].
8 *Ibid.*, p. 157.
9 David Knowles, *The Evolution of Medieval Thought* (New York: Vintage Books, 1962), p. 57.

CHAPTER VII

1 Frances Yates, *Giordano Bruno and the Hermetic Tradition* (London: Routledge & Kegan Paul, 1964). This work launched a pioneering area of academic inquiry that is still much debated. See also by the same author, "The Hermetic Tradition in Renaissance Science," in *Art, Science and History in the Renaissance*, ed. Charles Singleton (Baltimore: Johns Hopkins Press, 1968), pp. 255–74; *The Art of Memory* (Chicago: University of Chicago Press, 1966); *Theater of the World* (Chicago: University of Chicago Press, 1969); *The Rosicrucian Enlightenment* (London: Routledge & Kegan Paul, 1972). See also Robert S. Westman and J. E. McGuire, *Hermeticism and the Scientific Revolution* (Los Angeles: Clark Library, UCLA, 1977); and Brian Vickers, ed., *Occult and Scientific Mentalities in the Renaissance* (Cambridge: Cambridge University Press, 1984).
2 Yates, *Bruno*, pp. 1–19; D. P. Walker, *The Ancient Theology* (Ithaca, N.Y.: Cornell University Press, 1972), pp. 1–21.
3 Yates, *Bruno*, pp. 9–11.
4 Walker, p. 14.
5 Yates, *Bruno*, p. 14.
6 Walker, P. 20; Paul O. Kristeller, *The Philosophy of Marsilio Ficino*, tr. Virginia Conant (New York: Columbia University Press, 1943; reprint edition, Gloucester, Mass.: Peter Smith, 1964), pp. 26–7.
7 Kristeller, *Ficino*, pp. 22–3.
8 Walker, p. 14.
9 Kristeller, pp. 27–8, 322.
10 *Ibid.*, pp. 28–9.
11 Paul Oskar Kristeller, *Renaissance Thought and its Sources* (New York: Columbia University Press, 1979), pp. 57–8; R. I. Wallis, *Neo-Platonism* (New York: Charles Scribner's Sons, 1972), pp. 160–78.
12 Paul Oskar Kristeller, *Eight Philosophers of the Italian Renaissance* (Stanford: Stanford University Press, 1964), pp. 54–71; Ernst Cassirer, Paul O. Kristeller, John H. Randall, Jr., eds., *The Renaissance Philosophy of Man* (Chicago: University of Chicago Press, 1948), pp. 215–22.
13 Giovanni Pico della Mirandola, "Oration on the Dignity of Man," in Ernst Cassirer, *et al.*, eds., *The Renaissance Philosophy of Man*, cited above, pp. 249–50.
14 *Ibid.*, p. 251.

15 *Ibid.*, p. 252.
16 On the magical aspect of Renaissance Hermeticism, see D. P. Walker, *Spiritual and Demonic Magic from Ficino to Campanella* (London: The Warburg Institute, 1958; Kraus Reprint, 1969; University of Notre Dame Press, 1975).
17 This archetypal pattern has a long history in human affairs. See, for example, Mircea Eliade, *Rites and Symbols of Initiation* (New York: Harper & Row, 1965).
18 *Renaissance Philosophy*, cited above, pp. 247–8, 252.
19 These figures are studied as part of the pattern of transmitting the Cabalist tradition by Frances Yates in *The Occult Philosophy in the Elizabethan Age* (London: Routledge & Kegan Paul, 1979).
20 For example, Fludd refers to the *De arte cabalistica* eleven times in the *Mosaicall Philosophy*.
21 Peter J. Ammann, "The Musical Theory and Philosophy of Robert Fludd," JWCI, 30 (1967), 198–227.
22 E.g., *Mosaicall Philosophy*, p. 196 and many others. For other compendia he used, see Appendix.
23 See Walter Pagel, *Das medizinische Weltbild Paracelsus: seine Zusammenhange mit Neuplatonismus und Gnosis* (Wiesbaden: Franz Steiner Verlag, 1962); by the same author, *Paracelsus: An Introduction to Philosophical Medicine in the Era of the Renaissance* (Basel and New York: S. Karger, 1958); Allen Debus, *The Chemical Philosophy* (New York: Science History Publications, 1977).
24 Robert Fludd, *Mosaicall Philosophy* (London: Humphrey Moseley, 1659), p. 244.
25 Fludd owned a copy of Zetner's *Theatrum Chemicum*, which included the *Monas*; see Appendix.
26 On Dee, see I. R. F. Calder, "John Dee Studied as a Neoplatonist," unpublished M. A. Thesis, Warburg Institute, Univesity of London, 1962; Peter J. French, *John Dee: The World of an Elizabethan Magus* (London: Routledge & Kegan Paul, 1972); Gerald Suster, ed., *John Dee: Essential Readings* (Wellingborough, Northamptonshire: Aquarian Press, 1986); and Frances Yates' books listed in Bibliography.
27 E. E. Reynolds, *St. Thomas More* (New York: Doubleday Image Books, 1958), p. 21.
28 Ernst Cassirer, *The Platonic Renaissance in England*, tr. James Pettigrove (New Haven: Yale University Press, 1953; reprint edition, New York: Gordian Press, 1970), pp. 15–16.
29 Reynolds, p. 29.
30 Erasmus, *Erasmi Epistolae* 12 vols. (Oxford University Press, 1906–1958), X, p. 139 [letter 2750, to John Faber]. On More's Platonism, see Cassirer, *Platonic Renaissance, op. cit.*, pp. 22–4; Thomas I. White, "Pride and the Public Good: Thomas More's Use of Plato in *Utopia*," *Journal of the History of Philosophy* 20 (1982), 329–54; J. H. Hexter, "The Compositon of *Utopia*," in *The Complete Works of St. Thomas More* 4 (New Haven: Yale University Press, 1965), pp. xv–clxxxi.

NOTES

31 Mordechai Feingold, "The occult tradition in the English universities of the Renaissance: a reassessment," in Brian Vickers, ed., *Occult and Scientific Mentalities in the Renaissance, op. cit.*, pp. 73–94.

32 *Ibid.* Platonism in seventeenth-century England reached its peak with the Cambridge Platonists, led by Ralph Cudworth (1617–88) and Henry More (1614–87).

CHAPTER VIII

1 Fludd, "A Philosophical Key," Trinity College, Cambridge, Western MS. 1150 (0.2.46), fol. 21^r. Published edition by Allen Debus, *Robert Fludd and His Philosophical Key* (New York: Science History Publications, 1979), p. 76.

2 For an excellent survey of Fludd's illustrations with commentary, see Joscelyn Godwin, *Robert Fludd: Hermetic Philosopher and Surveyor of Two Worlds* (London: Thames & Hudson; Boulder: Shambala, 1979).

3 Fludd, *Mosaicall Philosophy* (London: Humphrey Moseley, 1659), p. i of preface.

4 *Ibid.*, p. ii.

5 *Ibid.*

6 *Ibid.*, p. 42.

7 *Ibid.*, p. 41.

8 *Ibid.*, p. 42.

9 *Ibid.*, p. 27.

10 *Ibid.*, p. 9.

11 *Ibid.*, p. 28.

12 See Allen Debus, "The Sun in the Universe of Robert Fludd," *Le Soleil à la Renaissance* (Brussels: Presses Universitaires de Bruxelles, 1965).

13 Fludd, *Monochordum Mundi* . . . (Frankfurt: de Bry, 1623).

14 *Mosaicall Philosophy*, p. 140, *et passim*.

15 "A Philosophical Key," fol. 104.

16 *Ibid.*, fol. 104^v.

17 *Ibid.*, fols 104–104^v; *Mosaicall Philosophy*, p. 44. Fludd does not give a specific reference to Plato or Augustine here.

18 *Ibid.*, fols. 104^v–105.

19 *Ibid.*, fols. 105^v–106.

20 *Ibid.*, fol. 106.

21 *Ibid.*, fol. 106^v. See Plato, *Timaeus*, 31b–32c. Although this theme was also used by the Hellenistic and Renaissance Neoplatonists, it is my belief that Fludd was following Plato directly.

22 *Ibid.*

23 *Ibid.*, fol. 108.

24 *Ibid.*

25 *Ibid.*, fol. 108^v. The musical harmonics of Fludd's universe have been thoroughly analyzed by P. J. Ammann (see Bibliography). On Fludd's sun, see Allen Debus, note 12.

26 "Philosophical Key," fols 109^v–110.

27 *Mosaicall Philosophy*, p. 151.
28 For a fascinating theory on the creation by Pythagorean-Platonic harmonics as tuning theory, see Ernest G. McClain, *The Myth of Invariance* (New York: Nicolas Hays, 1976) and also his *The Pythagorean Plato* (New York: Nicolas Hays, 1978). See also Kenneth S. Guthrie, compiler and translator, *The Pythagorean Sourcebook and Library* (Grand Rapids, Mich.: Phanes Press, 1987).
29 It was Fludd's notion that the acceptance of these three principles by Plato and Hermes proved they had read the Mosaic books and could therefore be considered divine and used as Christian sources. Since Aristotle did not accept these principles, according to Fludd, he was to be studiously avoided as unfit for Christian philosophers.
30 Fludd, *Mosaicall Philosophy*, p. 42. Fludd inserts this note: "Demogorgon signifieth the God of the earth and Universe, also the terrible God because he is greater than al the rest. This the Poets acknowledge to be he who hath created all things. Leon. Hebr. Dial:2." (i.e., Leo Hebraeus, or Jehuda ben Isaac Abarbel).
31 Here is inserted a note about this creature: "Litigium by con. Hebr: Dial. 2. Is sayed to be the first borne sonne of Chaos, haveing a crule and deformed countenance, who immediately after his birth stirred up debates and sought to fly upwards arrogantly, wherfore by Demogorgon he was cast downe into the bottom of darkness. Imagin therfore this was the prince of darkness."
32 "Philosophical Key," fols. 22–22^v.
33 *Ibid.*, 22^v–23.
34 *Ibid.*, fol. 23.
35 *Ibid.*
36 *Ibid.*, fols 23–23^v.
37 *Ibid.*, fol. 23^v.
38 *Ibid.*, fol. 24^v.
39 *Ibid.*, fols. 24^v–25.
40 *Ibid.*, fol. 26.
41 *Ibid.*, fols. 26–26^v.
42 *Ibid.*, fol. 27.
43 *Ibid.*, fols. 27–27^v.
44 *Ibid.*, fol. 27v.
45 *Ibid.*, fols. 29–29^v.
46 *Ibid.*, fol. 29^v.
47 Giovanni Pico della Mirandola, *Oration on the Dignity of Man*, trans. Elizabeth Forbes, in Ernst Carrirer, *et al.*, *The Renaissance Philosophy of Man* (Chicago: University of Chicago Press, 1971), p. 224.
48 "Philosophical Key," fol. 30.
49 Pico, *Oration*, p. 225.
50 "The Philosophical Key," fols. 38^v–39.
51 *Ibid.*, fols. 39^v–40.
52 *Ibid.*, fol. 5^v.
53 *Ibid.*, fol. 40.
54 *Ibid.*

55 Pico, *Oration*, p. 228.
56 Edgar Wind, *Pagan Mysteries in the Renaissance* (New York: Norton, 1968), pp. 20–1. Wind quotes Augustine from *Retractiones* I, xiii.
57 Pico, *Oration*, p. 247.
58 *Ibid.*, pp. 248–9.
59 *Ibid.*, p. 249.
60 For further details of the weapon-salve, see Allen Debus, *The English Paracelsians* (New York: Franklin Watts, 1966), p. 121.
61 "Philosophical Key," fol. 17.
62 *Ibid.*, fol. 44.
63 *Ibid.*
64 *Ibid.*, fol. 44^v.
65 *Ibid.*, fols. 44^v–45.
66 *Ibid.*, fol. 45.
67 *Ibid.*, fols. 45^v–46.
68 *Ibid.*, fols. 46–46^v.
69 *Ibid.*, fol. 45.
70 Fludd, *Anatomiae Amphitheatrum* (Frankfurt: de Bry, 1623).
71 Fludd, *Mosaicall Philosophy* (London: Humphrey Moseley, 1659); *Philosophia Moysaica* (Gouda: Petrus Rammazenius, 1638).
72 *Mosaicall Philosophy*, p. 4.
73 *Ibid.*, p. 6.
74 The historical background of this instrument is given in an article by F. Sherwood Taylor, "The Origin of the Thermometer," *Annals of Science* 5 (1942), pp. 129–56.
75 *Mosaicall Philosophy*, p. 7.
76 *Ibid.*, p. 9.
77 *Ibid.*, p. 35.
78 *Ibid.*
79 *Ibid.*
80 *Ibid.*
81 *Ibid.*, p. 242.
82 *Ibid.*, p. 244.
83 William Munk, *Roll of the Royal College of Physicians of London*, 2 vols. (London: Longman, *et al*, 1861), p. 73.
84 *Mosaicall Philosophy*, pp. 202–3.
85 *Ibid.*, p. 38.
86 Fludd felt this was proven by his Weather-Glass experiment discussed above.
87 Although Fludd's sources are sometimes difficult to trace, the system described in this section appears to follow Ficino's combination of medieval Christian hierarchies with Neoplatonist sources, including Hermes Trismegistus (see Kristeller, *The Philosophy of Marsilio Ficino*, pp. 75 ff.) and Pico's addition of the Cabala to the entire schema. See Pico's "Oration," previously cited, pp. 226–8 and p. 251. On Fludd's use of Cabala, see A. E. Waite, *The Holy Kabbalah* (New Hyde Park, NY: University Books, 1960), pp. 467–9.
88 *Mosaicall Philosophy*, p. 59.

89 *Ibid.*, p. 60.
90 *Ibid.*, p. 124.
91 *Ibid.*, p. 178.
92 *Ibid.*
93 *Ibid.*
94 *Ibid.*, pp. 180–1.
95 *Ibid.*, pp. 181–2.
96 *Ibid.*, p. 182.
97 *Ibid.*
98 *Ibid.*
99 *Ibid.* No one has investigated Fludd's astrology, nor can it be
    examined at length here. Again Fludd is attempting to synthesize
    obvious Cabalistic doctrine with the Neoplatonic-Hermetic. His
    views seem similar to those of Paracelsus, but Paracelsus was
    operating much in the same tradition (see Walter Pagel, *Paracelsus*,
    cited in Bibliography) and his writings are often far from clear.
    Fludd's notions would also seem to be compatible with those of
    Ficino (even though both Ficino's and Pico's genuine feelings about
    astral influences have been the subject of much debate: see Kristeller,
    Ficino, previously cited, pp. 310–12 *et passim*; and D. P. Walker,
    Spiritual and Demonic Magic from Ficino to Campanella, pp. 54–9 *et
    passim*). By distinguishing between "vulgar" astrology and his own
    kind, Fludd apparently meant that the former was a type that
    assumed the physical astral body to exert some kind of influence over
    men from which they could not escape; these were the charts sold by
    the common back-alley astrologer. On the other hand, Fludd's charts
    showed occult spiritual influences at work which *could* be overcome by
    the application of a remedy which restored a lacking benevolent
    divine force. His attitude implies both a free will and man's ability to
    triumph over or manipulate nature if he attains an understanding of
    the true but occult forces operating behind sensually perceived
    phenomena. This view seems to me to be quite similar to the views of
    Ficino, Pico and Paracelsus.
100 *Mosaicall Philosophy*, p. 41.
101 *Ibid.*
102 *Ibid.*, p. 53.
103 *Ibid.*, p. 55. See the Weather-Glass experiment in Section III.
104 *Ibid.*, p. 190.
105 *Ibid.*, p. 189.
106 "Philosophical Key," fols. 56–56ᵛ.
107 *Ibid.*, fol. 56.
108*Ibid.*

CHAPTER IX

1 Thomas Fuller, *The Worthies of England* (London, 1662; also London:
   Rivington, *et al.* 1811) I, p. 503.

Anthony Wood, *Athenae Oxonienses* (London, 1691; also London: Rivington, et al. 1815) II, p. 618.

2 James Brown Craven, *Doctor Robert Fludd: The English Rosicrucian* (Kirkwall: Wm. Peace & Sons, 1902), pp. 14 and 40.

3 Arthur Edward Waite, *The Brotherhood of the Rosy Cross* (London, 1924; New Hyde Park, NY: University Books, n.d.) p. 307.

4 Frances Yates, *The Rosicrucian Enlightenment* (London and Boston: Routledge & Kegan Paul, 1979), p. 179.

5 For some modern scholarship which deals with the Rosicrucian problem, see John Warwick Montgomery, *Cross and Crucible* (The Hague: Martinus Nijhoff, 1973); Frances Yates, *The Rosicrucian Enlightenment*, cited above; A. E. Waite, *The Real History of the Rosicrucians* (London: Redway, 1887); by the same author, *The Brotherhood of the Rosy Cross*, cited above; the Montgomery book contains a very valuable annotated bibliography.

6 By Michael Maier. See Waite, *Brotherhood*, p. 117.

7 Waite, *History*, p. 37; an English translation of the "Universal Reformation" is in this volume, pp. 36–63.

8 *Ibid.*, p. 63.

9 The texts used are the English translations by Thomas Vaughn of 1652, as edited by Frances Yates in *The Rosicrucian Enlightenment*, pp. 238–60, and in Waite, *History*, pp. 64–98.

10 Yates, *Enlightenment*, p. 238.

11 *Ibid.*, p. 239.

12 *Ibid.*, p. 241.

13 *Ibid.*

14 *Ibid.*, p. 243.

15 *Ibid.*, p. 241.

16 *Ibid.*, p. 250.

17 See *ibid.*, p. 39.

18 Waite, *History*, pp. 85–6.

19 *Ibid.*, p. 86.

20 Montgomery, *Cross and Crucible*, p. 193.

21 On Libavius, see Lynn Thorndike, *A History of Magic and Experimental Science*. 8 vols. (New York: Columbia University Press, 1923–1958), VI, 238–53; Allen Debus, *The Chemical Philosophy* (New York: Science History Publications, 1977), pp. 171–86; Waite, *Brotherhood*, pp. 236–40; Yates, *Enlightenment*, pp. 51–3.

22 *Apologia Compendiaria, Fraternitatem de Rosea Cruce suspicionis et infamiae maculis aspersam, veritatis quasi Fluctibus abluens et abstergens* (Leyden: Godfrey Basson, 1616).

23 Leyden: Godfrey Basson, 1617. On Basson, see J. A. van Dorsten, *Thomas Basson, 1555–1613, English Printer at Leiden* (Leiden: Sir Thomas Browne Institute, 1961).

24 *Tractatus Apologeticus*, p. 195.

25 Theodor de Bry (1528–98), originally a Walloon copperplate engraver in the Spanish Netherlands, established the business in Frankfurt in 1590. Upon his death, it was taken over by his two sons, Johann

Theodore (1561–1623) and Johann Israel (?–1611). Because of religious struggles they moved to Oppenheim in the Palatinate in 1609, where Johann Israel died two years later. In 1617, the talented engraver Matthieu Merian (1593–1650) joined the firm and married a daughter, Maria Magdalena de Bry, one year later. (Merian engraved the Fludd portrait of 1626 in the *Philosophia Sacra*.) In 1619 de Bry moved back to Frankfurt, where he died in 1623, after which the business was taken over by Merian. Two years later, the Englishman William Fitzer joined the firm as a partner, married another daughter, Susanna, and soon after had his own publishing business in Frankfurt. Moritz Sondheim, "Die de Bry, Matthäus Merian und Wilhelm Fitzer," *Philobiblon* 6 (1933), pp. 8–34; E. Weil, "Wilhelm Fitzer, The Publisher of Harvey's *De motu cordis*, 1628," *The Library*, 4th series, 24 (1944), pp. 142–64; Yates, *Enlightenment*, pp. 70–90.

26 "Declaratio Brevis," fol. 9^v.

27 Published with an introduction by Allen Debus, *Robert Fludd and His Philosophical Key* (New York: Science History Publications, 1979).

28 "Philosophical Key," fols. 5^v–6^r; Debus, p. 68.

29 *Ibid.*, fols. 7^r–7^v; Debus, pp. 68–9.

30 *Ibid.*, fols. 14^v–15^v; Debus, pp. 72–3.

31 *Ibid.*, fols. 15^v–15^v; Debus, p. 73.

32 *Ibid.*, fol. 70^r; Debus, p. 109.

33 *Ibid.*, fol. 69^v; Debus, p. 109.

34 James Brown Craven, *Count Michael Maier* (Kirkwall: Wm. Peace & Son, 1910), pp. 108; Waite, *Brotherhood*, pp. 310–39.

35 DNB XV, p. 35.

36 Donald S. Pady, "Sir William Paddy, M.D. (1554–1634)," *Medical History* 18 (1974), p. 77.

37 *Ibid.*

38 C. H. Josten, "Robert Fludd's 'Philosophical Key' and His Alchemical Experiment on Wheat," *Ambix* 11 (1963), pp. 1–2. A reproduction of Fludd's entry in Morsius' book with the annotations by Morsius of Fludd's books is in Heinrich Schneider, *Joachim Morsius und Sein Kreis* (Lübeck: Otto-Quitzow-Verlag, 1929), p. 23. It is interesting to note that Morsius lists a "Lib. de Vita, Morte et Resur." as one of Fludd's works, i.e., the *Tractatus Theologo-Philosophicus* of 1617.

39 Frankfurt: Lucas Jennis, 1617.

40 *Ibid.*

41 Frankfurt: Lucas Jennis, 1618.

42 A bibliographic survey of Maier's publications may be found in Craven, *Count Michael Maier*, cited above. See also Waite, *Brotherhood*, pp. 310–39.

43 *Dr. Fludds Answer unto M. Foster* (London: Nathaniel Butler, 1631), pp. 21–2.

44 Yates, *Enlightenment*, pp. 114–16.

45 Joachim Frizius, *Summum Bonum* (n. p. [Frankfurt]: Officina Bryana, 1629), p. 2.

NOTES

46 *Ibid.*, pp. 51–2. An old English translation of the letter is printed in Waite, *Real History*, pp. 296–300.
47 *Clavis Philosophiae et Alchemiae Fluddanae* (Frankfurt: William Fitzer, 1633), p. 20.
48 *Ibid.*, p. 26.
49 *Ibid.*, p. 20.
50 The *Clavis* was finished by 1631, since Fludd mentions it in *Dr. Fludds Answer*, but not published until 1633.
51 *Clavis*, pp. 22–3; *Summum Bonum*, p. 51.
52 *Dr. Fludds Answer*, p. 23.
53 Yates, *Rosicrucian Enlightenment*, pp. xi–xiv.
54 Robert Westman, "Magical Reform and Astronomical Reform: The Yates Thesis Reconsidered," in Robert S. Westman and J. E. McGuire, *Hermeticism and the Scientific Revolution* (Los Angeles: W. A. Clark Library, UCLA, 1977).
55 Montgomery, *Cross and Crucible*, pp. 210 and 232 ff.
56 *Ibid.*, pp. 201–6; Waite, *Brotherhood*, pp. 39–54; Yates, *Enlightenment*, pp. 33–5.
57 Montgomery, p. 232.
58 Hermann Schlenz, "Goldmachen und Goldmacher am hessischen Hofe," *Deutsche Geschictsblätter* 11 (1910), p. 310.
59 *Ibid.*, pp. 301–27; Christoph von Rommel, *Geschichte von Hessen* 9 vols. (Cassel: Hampe, *et al.*, 1820–1853).
60 Rommel, 6(2), 383 ff.

CHAPTER X

1 Jean-Jacques Boissard, *Bibliotheca sive thesaurus virtutis et Gloriae* (Francofurti: Artificiossissime in aes incissae a loan. Theodor. de Bry, Suptibus Guiliemi Fitzeri, 1628–30), Part 2, p. 198.
2 See discussion of Wolfgang Pauli in Chapter V.
3 William Irwin Thompson, *Darkness and Scattered Light* (Garden City, N.J: Anchor Books, 1978), pp. 20, 122.
4 Charlotte Fell Smith, *John Dee (1527–1608)* (London: Constable, 1909); Richard Deacon, *John Dee* (London: Frederick Muller, 1968); Peter French, *John Dee* (London: Routledge & Kegan Paul, 1972).
5 A. L. Rowse, *Simon Forman* (New York: Scribner's, 1974).
6 Smith, p. 18; French, p. 6.
7 French, pp. 40–61.
8 John Dee, *A True and Faithful relation of what passed for many years between Dr. John Dee and some spirits*, preface by Meric Casaubon (London: Maxwell for T. Garthwait, 1659), p. 38 of the preface.
9 Rowse, pp. 259–61.
10 William Lilly, *History of His Life and Times* (London: Charles Baldwyn, 1822), pp. 35–6.
11 Anthony Wood, *Athenae Oxonienses*, ed Bliss (London: Rivington, et al.) Vol II, col 100.

12 Bacon, *New Organon*, Book Two, Aphorism 38. See also F. Sherwood Taylor, "The Origin of the Thermometer," *Annals of Science*, 5 (1942), pp. 143–50.

13 Bacon, *De Augmentis Scientiarum* in John Robertson, ed., *The Philosophical Works of Francis Bacon* (London: Routledge, 1905), p. 497.

14 See Chapter VII.

15 Bacon, *Novum Organum* in Robertson, op. cit., p. 272. The charge in this passage about a "heretical religion" in such a philosophy was a serious one. Allen Debus has suggested that Bacon's denouncing of Fludd's philosophy may have been the basis for the King's examination of Fludd, resulting in the writing of the *Declaratio Brevis* and the *Philosophical Key*, but there is no evidence for this. See Debus' introduction to the *Philosophical Key*, cited in Bibliography.

16 Before Bacon published his *New Organon* in 1620, Fludd had completed all that was ever published of his multi-tomed *Utriusque cosmi . . . historia*, a history of the macrocosm and microcosm, between 1617 and 1626 (see Bibliography), which included an exposition of his philosophy.

17 Bacon, *New Organon*, Book One, Aphorism V.

18 See, for example, d'Alembert's statements about Bacon in the "Preliminary Discourse" to the *Encyclopedia*.

19 Fludd's battle with Kepler is discussed in detail in Chapter V.

20 W. Foster, *Hoplocrisma Spongus, or a Sponge to wipe away the weapon-salve* (London: Thomas Cotes, 1631).

21 Dr. Baldwyn Hamey the younger, "Bustorum aliquot Reliquae," Royal College of Physicians MS 149; British Museum Sloane MS 2149.

22 Thomas Fuller, *The History of the Worthies of England* (London: Rivington, *et al.*, 1811), vol. I, pp. 503–4.

23 Elias Ashmole, *Theatrum Chemicum Britaññicum* (London: 1652). Reprinted with an introduction by Allen Debus (New York and London: Johnson Reprint, 1967) (Sources of Science Series, no. 39), p. 460.

24 In 1677, one voice was still being raised in Fludd's defense: John Webster says in his *The Displaying of Suposed Witchcraft* (London: Jonas Moore, 1677), p. 9:

> Our Countryman Dr. Fludd, a man acquainted with all kinds of Learning and one of the most Christian Philosophers that ever writ, yet wanted not those snarling Animals, such as Mersennius, Lanovius, Foster, and Gassendus, as also our Causaubon (as mad as any) to accuse him vainly and falsely of Diabolical Magick, from which the strength of his own Pen and Arguments did discharge him without possibility of replies.

See Allen Debus, *Science and Education in the Seventeenth Century: The Webster-Ward Debate* (New York: American Elsevier, 1970).

25 Sir William Temple, *Works*, vol. III (London: J. Clarke, *et al.*, 1792). p. 452.

26 Anthony Wood, *Athenae Oxonienses*, ed. Philip Bliss, vol. II (London: J. Rivington *et al.*, 1815), cols 617–22.

27 Rev. J. Granger, ed., *A Biographical History of England*, vol. II (London: T. Davies, *et al.,* 1775), pp. 3–4.

28 John Aiken, *et al.*, eds., *General Biography*, vol. IV (London: J. Johnson, *et al.*, 1803), p. 142.

29 William Munk, *Roll of the Royal College of Physicians*, vol. I (London: Longman, Green, *et al.*, 1861), pp. 140–3.

30 *Ibid.*, p. 142.

31 Gordon Wolstenholme, ed. *The Royal College of Physicians: Portraits* (London: J. & A. Churchill, 1964), pp. 162–3.

32 A. E. Waite, *The Real History of the Rosicrucians* (London: George Redway, 1887), pp. 283–307.

33 See, for example, the DNB article on Fludd.

34 J. B. Craven, *Doctor Robert Fludd* (Kirkwall, Orkney Islands: William Peace & Son, 1902).

35 A recent work by Serge Hutin is *Robert Fludd (1574–1637), Alchemiste et Philosophe Rosicrucien*, Collection "Alchimie et Alchimistes" No. VIII (Paris: Omnium Litteraire, 1971). This is basically Hutin's 1951 thesis for his *Elève diplômé*. It uncritically follows various occultist works, contains a number of errors and bad references, and does not take into account any of the modern scholarship on Fludd.

36 See Bibliography.

37 Frances Yates, *Giordano Bruno and the Hermetic Tradition* (Chicago: University of Chicago Press, 1964).

CHRONOLOGY

1574 January. Born at Milgate House, Bearsted, Kent. Baptized in Bearsted Parish Church 17 January.

1592 25 January. Mother, Elizabeth Andrews Fludd, dies at Milgate House.
10 November. Enters St John's College, Oxford.

1596 3 February. Receives B.A. from St John's.

1596–8 Studies for M.A. at St John's. Writes treatise on music.

1598 8 July. Receives M.A. from St John's, says he is going overseas.

1598–1604 Travels in France, Spain, Italy and Germany. Tutors Duke of Guise and his brother. Writes treatises on arithmetic, geometry, perspective, military arts, art of memory, geomancy, motion, astrology.

1604 or 1605 Enters Christ Church, Oxford.

1605 16 May. Receives M.B. and M.D., licenced to practice medicine.
8 November. First examination by College of Physicians in London to practice medicine there.

1606 7 February. Examined by College, given permission to practice.
2 May. Questioned by College about allegations of arrogance concerning supremacy of chemical medicines over Galenical.
July. Travels to France to confer with colleagues from Italy and France.

1607 30 May. Father, Sir Thomas Fludd, dies at Milgate House.
1 August, 9 October, 22 December. Further examined by the College.

1608 21 March. Offends Censors of College of Physicians by examination replies. Candidacy for Fellowship in College revoked.
25 June. Readmitted as candidate for Fellowship in College of Physicians.

1609 20 September. Admitted as a Fellow of College of Physicians of London.

c. 1610 Completes MS. of *History of the Macrocosm*. Read by John Selden, medical colleague Dr Richard Andrewes and others.

1614 John Selden praises Fludd's medical skill in his *Titles of Honor*.

Fama Fraternitatis of the Order of the Rosy Cross published in Germany.

1615 *Confessio Fraternitatis R. C.* published in Germany.

Andreas Libavius attacks the Fraternity in *Analysis Confessionis Fraternitatis De Rosea Cruce*.

1616 Fludd replies to Libavius with *Apologia Compendiaria*. The brief Apology includes an outline for a longer work and a letter to the Fraternity.

1617 The longer defense, *Tractatus apologeticus integritatem Societatis de Rosea Cruce defendens*, published in Leyden.

The *Tractatus Theologo-Philosophicus . . . de Vita, Morte et Resurrectione* published in Oppenheim by de Bry.

Also from the de Bry press appeared the first part of Fludd's *magnum opus: Utriusque cosmi majoris scilicet et minoris metaphysica, physica, atque technica historia* (Technical, Physical and Metaphysical History of the Macrocosm and Microcosm). Contains Volume I, *History of the Macrocosm*, Tractate I. Dedicated to God and James I.

1618 *De Naturae Simia* (The Ape of Nature) printed by de Bry in Oppenheim.

Tractate II of the *History of the Macrocosm*.

12 May. Two holders of the monopoly patent to make steel in England complain to the Privy Council that Fludd is making steel illegally in his chemical laboratory.

Fludd elected a Censor of the College of Physicians.

Called before James I to defend his Apology and Macrocosm History.

Interview gains Fludd the favor and patronage of the King. Writes "Declaratio Brevis" at suggestion of James.

1619 Volume II, *History of the Microcosm*, Tractate I, published by de Bry in Oppenheim: *Tomus Secundus, de supernaturali, naturali, praeternaturali et contranaturali microcosmi historia*.

Fludd writes "A Philosophical Key" as sequel to the "Declaratio Brevis".

Johann Kepler publishes his *Harmonices mundi*. Includes a lengthy appendix attacking Fludd's version of the Neoplatonic harmonies of the universe.

1620 30 May. James I charges Privy Council to consider

Fludd's petition to be granted a patent to make steel.

27 June. Fludd gives public anatomy lecture at College of Physicians.

27 September. The Privy Council grants Fludd a patent to make steel after considering testimony of its superiority.

1621 Another part of the unfinished Microcosm History published by de Bry in Frankfurt: *Tomi secundi tractatus secundus, de praeternaturali utriusque mundi historia.*

The *Veritatis Proscenium* also published in Frankfurt by de Bry. Replies to Kepler's attack.

1622 Kepler replies to Fludd in his *Pro suo opere harmonices mundi apologia.*

1623 *Anatomiae Amphitheatrum*, Fludd's mystical anatomy, published in Frankfurt by de Bry. Dedicated to John Thornborough, Bishop of Worcester. At the end is the *Monochordum Mundi*, dated 1621, Fludd's final revised universal harmonic scheme and last reply to Kepler.

Marin Mersenne, the French mechanist, attacks Fludd's mystical philosophy and science in his *Quaestiones celeberrimae in Genesim.*

1625 Death of James I, accession of Charles I.

Visited Bishop of Worcester at Hartlebury Castle. Viewed a nearby lightning strike site with Mr Finch and the Bishop's son, Sir Thomas Thornborough.

1626 *Philosophia sacra et vere Christiana seu Meteorologia Cosmica* published in Frankfurt by Officina Bryana. Dedicated to John Williams, Bishop of Lincoln.

1627 Elected a Censor of the College of Physicians.

20 July. Inspects alum works with William Harvey and six others from the College at the order of the Privy Council.

1629 8 June. Grant by Charles I to Fludd and his heirs of a "messuage and lands" in Suffolk.

Replies to attacks by Mersenne in *Sophie cum moria certamen.* Bound with it is the *Summum Bonum*, dealing with the Rosicrucians, by Joachim Frizius.

First part of the *Medicina Catholica* published by Fitzer in Frankfurt.

Dedicated to Sir William Paddy.

1630(?) *Pulsus*, the second part of the *Medicina Catholica*, published in Frankfurt, the first printed work to agree with William Harvey's circulation of the blood theory.

Pierre Gassendi publishes his examination of Fludd's works, done at Mersenne's request, in his *Epistolica*.

Donates twenty-four books to Jesus College, Oxford.

1631 The third part of the Catholic Medicine, *Integrum Morborum Mysterium*, dedicated to George Abbot, Archbishop of Canterbury, and the fourth part, *KAΘOLIKON Medicorum KATOΠTRON*, dedicated to Sir Robert Bruce Cotton, published in Frankfurt by William Fitzer.

William Foster, an English cleric, attacks Fludd's views on the weapon-salve as diabolical. Fludd replies with *Doctor Fludds Answer unto M. Foster*.

1633 Elected a Censor of the College of Physicians.

The *Clavis Philosophiae Et Alchymiae Fluddanae*, the final refutation of Mersenne and Gassendi, published in Frankfurt by Fitzer.

1634 12 June. Sworn in as a brother in the Barber-Surgeon's Company.

Elected a Censor of the College of Physicians.

1637 8 September. Dies in his house in London.

Buried in Bearsted Parish Church.

Leaves Manuscript published as *Philosophia Moysaica* in Gouda, 1638, and *Mosaicall Philosophy*, London, 1659.

ROBERT FLUDD'S "DECLARATIO BREVIS" TO JAMES I

This unpublished Latin manuscript in the British Library (Royal MS. 12 C ii) was written in late 1618 or early 1619 as a follow-up to Robert Fludd's interview with James I, where he defended his *Tractatus Apologeticus* and the *History of the Macrocosm*, which was dedicated to the King. As a result of the interview, Fludd gained the patronage and good will of the monarch until his death in 1625. The "Declaratio Brevis" was followed by another unpublished manuscript, in English, "A Philosophical Key," which was meant to further explain his philosophy, as discussed in Chapter VI. The Latin text, the English translation and an introduction by William H. Huffman were first published in *Ambix* 25 (1978), pp. 69–92. The manuscript was transcribed and edited by William H. Huffman and Robert A. Seelinger, Jr, and the translation was by Robert A. Seelinger, Jr, with the editorial collaboration of William H. Huffman; the latter also wrote the notes to the text. The translation is reprinted here with the kind permission of the Society for the History of Alchemy and Chemistry.

A brief declaration
dedicated to
the most serene and powerful Prince
and Lord, Lord JAMES, King of
Great Britain, France and Ireland,
Defender of the Faith

In which the sincere intention of a certain publication
is explained very clearly to your Royal Majesty
by the present author, Robert Fludd,
Esquire and Doctor of Medicine,
the most loyal subject of
your Royal Majesty.

Behold, most serene King, that I have, as I hope, brought Your Majesty's auspicious mandate to its utmost conclusion; I have composed, on your gracious suggestion, a Declaration, not an Apology; moreover, we pray beforehand for forgiveness if I, the Author of this narration, am guilty of any troublesome error or unpolished style. For who, understanding the depth of Your Majesty's judgment, would not be disinclined to dedicate his books or writings to him [if they contain errors] or [subject them] to his scrutiny, by whose eyes, as if they were the eyes of a lynx, he examines every corner of a writer or [1^v] speaker, and immediately discovers those errors in the works of others which are not visible to or even perceived by the ample talents of the individual authors themselves. But in fact, Your Majesty orders, and my sense of obedience urges that something be published for the eyes of Your Majesty, whereby the face of truth will be disclosed with the veil of doubt removed, and thereby I may obtain the most desired thanks of Your Majesty. But my zeal towards Your Majesty urges me to persuade soothingly the receptive ears of Your Majesty with the calm whispering of truth. First of all, I shall begin to unfold to Your Majesty the line of reasoning of my published *Tractatus Apologeticus* so that henceforth all fancy of religious innovation or suspicion of heresy may be banished forever, and that the just intention of my mind in the publication of this work may be more easily and clearly known to Your Majesty. Therefore, in the first place, Your Majesty, it will appear most evidently, unless I am mistaken, that my *Tractatus Apologeticus* clearly does not deal with religious innovation, nor does it share even an iota of any heresy, inasmuch as I, the author of that work, have steadfastly adhered to this reformed religion (which is now the custom among us) from my infancy, and indeed almost from the time I lay at the breast of my nurse in England at the very beginning of my life and right up to this day; and I acknowledge and confess in the presence of God and Your Majesty, from the very bottom of my heart, that I remain a perfectly chaste man. Next, it is well known that my Apology, defending the Brothers of the Rosy Cross against the attacks of D. Libavius, first of all pertains particularly to the impediments of the Arts, which are in a state of decline, and the method of reviving them; and then afterwards considers [2^r] the wondrous qualities of Art and Nature with philosophical arguments and by frequently using the affirmations of the ancients. Furthermore, this school of Philosophers is acknowledged even by the Germans, Catholics as well as Lutherans (among whom the

Brothers are said to live), to embrace firmly the Calvinist Religion, just as it is rightful to infer from these lines extracted from a certain letter sent by a friend of mine from Frankfurt:

> The Fraternity of the Rosy Cross has been attacked by so many German writers, and clearly it is rejected by them, while they think the Theosophy of the Brothers is the Theology of the Calvinists, etc.
>
> Your most loyal friend,
Justus Helt[1]

Besides, even the Brothers themselves confirm in their Confession that they profess the reformed religion of Germany. From this, therefore, it is sufficiently evident to Your Majesty that neither the motive of innovating religion, nor the affection for some heresy, influenced me to publish the *Tractatus Apologeticus* and to have it distributed publicly. Consequently, concerning the reason that I published this *Tractatus* for the favor of the Fraternity, I will openly set forth to the just eyes of Your Majesty, the integrity of my [2ᵛ] heart and the truth of the matter itself, and with the number of words reduced to a minimum, I will clearly show it here in such a way that Your Majesty will not have any further doubts about my faithfulness towards God and Your Majesty and to the fatherland. There are two spurs of the mind exciting me to an understanding of and familiarity with these celebrated Philosophers: the revelation of the true basis of natural philosophy, commonly unknown to this day, and the discovery of the profound secret of medicine, celebrated so much with praises by Theography and Philosophy; this Society of the Rosy Cross professes to possess these two gifts of God and Nature (His servant) in the following propositions:

Proposition I

The true philosophy, commonly thought of as new, which destroys the old, is the head, the sum, the foundation, and the embracer of all Disciplines, Sciences, and Arts. This true philosophy, if we contemplate our world, will contain much of Theology and Medicine, but little of Jurisprudence; it will diligently investigate heaven and earth, and will sufficiently, by its images, explore, examine and depict Man, who is unique.

[3ʳ] Proposition II

We are able to show certain modest truths and things that are

211

useful to our country by which its various illnesses can be cured. These truths are not to be divulged in a common manner, which is uncertain and inconstant, but in a new way, unknown to the world, which is most certain and infallible.

And so these companion propositions have been the two lamps for my mind, which is most avid for knowledge, and, as it were, have been those two blessed luminaries Castor and Pollux, by whose sparks a great desire has been kindled in me to comprehend these men, who are most eminent in character and in their promises. And indeed (in my opinion) these propositions are sufficient for any curious spirit worthy of being inflamed by the fire of understanding, which spirit in times past has experienced the leprous notions of Philosophy and Medicine and their inconstancy. Moreover, if Your Majesty will deem it worthy to read through the lives of all the most distinguished philosophers and most skilled physicians (namely, among the philosophers, the life of Plato, Pythagoras, Thales [3ᵛ] of Melissus, Aristotle, Anaxagoras, Empedocles, Orpheus, Apollonius of Tyana, Hermes Trismegistus and many others) you will surely discover that all of them, for the sake of learning and erudition and to become participants in the mysteries of the divine philosophy, made long and laborious journeys through almost all of the learned world, in the manner of pilgrims, for the purpose of visiting the wise men of Ethiopia, examining the mysteries of the Egyptian High Priests, and for pursuing the enigmas of the hieroglyphics as well as the secrets of the inscriptions and carvings on the Pyramids of Memphis. And they even journeyed to oriental regions of the world, there to become thoroughly learned in the doctrine of natural things from the Magi or wise men of Babylonia, Persia and India. Whereby it has come about that by acquiring the immortal title of divinity from some of these sages, they are called Divine even by the Christians, since they acknowledge the Trinity not only of God, but also of the uncreated Persons. In this number, Plato and Hermes are counted in particular. Thus also Your Majesty will discover that among the physicians, Apollo, Aesculapius, Chiron the Centaur, Hippocrates of Cos, Chrysippus of Sicily, Aristratus of Macedonia, Euperices of Trinacria, Herosilus of Rhodes, Galen and many other men who were most learned and excellent in medicine also travelled through many nations of the world for the sake of acquiring experience in philosphy and medicine. They traversed all corners of the known world to discover the hidden mysteries of philosophy and [4^r] medicine by searching and by wearisome exertion; and above all, they visited the celebrated temple of Diana of Ephesus so that they might behold the medical records preserved therein. Whence at last they had become provided with knowledge, enriched and ennobled from their experience, they returned to their homeland and were held in the greatest esteem and honor by their peers because of the miracles they provided. And it also happened that some of them (upon whom people bestowed divine honors and worship, and for whom they set up graven images and statues, namely Apollo and Aesculapius) were regarded as gods by the pagan world. Similarly, Euperices and Herosilus were regarded by their peers as demi-gods. Thus also Hippocrates and Chrysippus acquired from the Greeks the highest honors as sanctified ones and saviors of the body. And, although this pagan superstition of those times must be regarded by us Christians as abominable, nevertheless we are never able to admire and praise highly enough those remarkable gifts and arcane mysteries of these distinguished physicians who were able to arouse such notable admiration even in the hearts and minds of the uncivilized and barbarian people. Allow me to conclude the first part of the present declaration with, as it were, one word, inasmuch as I am conscious of how precious each moment of time is for Your Most Serene [4^v] Majesty, having in your hand so many matters of great importance; and I fear lest a discourse, wearisome to you, be longer than this vile matter; but I say that many wondrous secrets lurk in the repository of nature, which not even to this day have been revealed to us. And since the honorable and lawful inquiry into the mysteries of nature has as yet by no age ever been prohibited or interdicted for the Philosopher and Physician, I, who profess to be a Philosopher and Physician (although not the most distinguished), have decided to offer this brief Declaration of mine to Your Most Venerable Majesty (even if I had not been urged to do so by your gracious suggestion) so that I may give satisfaction to Your Majesty concerning the reasoning of my previously published *Tractatus Apologeticus* and thus, after the good grace and favor of Your Majesty have been obtained, I may live among my Muses happily, cheerfully and free from envy, filled at last with unexpected joy.

In fact, concerning that which is important in my *Macrocosm*, I could write a very large volume (which I know would be wearisome for Your Majesty to read) in defense of each part of it. But among all other things here I will disclose the reason why I have dedicated my book to Your Majesty. I acknowledge that I dedicated this book especially to God [5^r] my Creator, because it is through Him that we act on this earth, breathe, live and possess all things which we enjoy in this life. It is certainly thus that whatever good is begun or completed by us must be particularly due to Him. Then in the second place I have dedicated this book to Your Majesty because immediately after God I acknowledge that the zeal of men and all their efforts ought to proceed to the honor of Your Majesty. Consequently, I have considered that it was permitted for me to depend upon your regal grace and gentleness to such a degree that in return for so many late-night studies or so much work I expected above all the grace and favorable view of Your Majesty. Moreover, since from a letter sent by a certain friend I realized that there was some controversy between the individual to whom I entrusted this volume in England[2] and the engraver and printer[3] concerning the dedication of my work. While the former endeavored to assign the honor of my book and labor to the Landgrave of Hesse, the latter individuals in fact endeavored to assign it to the Count Palatine, their own prince, and (as I have here some witnesses to this matter) at last I was compelled to transmit, unexpectedly, that twofold dedication, namely to God and to you, my King, so that I might absolutely prohibit them from assigning these works of mine to any mortal except to my King alone, to whom I acknowledge I owe what is mine. And this is the reason that I, a fosterling, [5^v] have chosen you alone before any foreign prince as my Maecenas and patron, not by any presumption, but induced by a love of Your Majesty, who is far better deserving than those who would have been substituted for you. But so that Your Majesty may realize that I have not done this rashly or indiscreetly, it will be well for Your Majesty to know that I received several letters from Germany informing me that men of letters, particularly of that country, and the learned of every profession, both Papist and Lutheran as well as Calvinist, praised far beyond my merits this volume of mine and seem to approve of my works unanimously. Whence considering that no one in the entire world was more worthy of the honor and dedication of these labors of mine, which have gained so much approval from the learned, than Your Majesty, my own King, who is quite distinguished in both sciences and letters; to him especially, before all others, I have willingly and with pleasure dedicated these works of mine. And although it may seem foul and indecent to praise oneself and commend one's own works, nevertheless I humbly entreat Your Majesty not to impute to me the charge of vainglory if by the testimony of foreigners I should protect my reputation, which is at stake in this work of mine – the *Macrocosm* – before Your Majesty, and should defend the sincerity of my writings. Therefore, for the sake of brevity, [6ʳ] I thought it worthy to place before the eyes of Your Majesty only the relevant portions of some of the letters sent to me from foreign countries lest the testimony of those previously mentioned by me seem to be made up or worthless. Therefore, first of all in supplication I so entreat Your Majesty not to reject the information that the engraver-printer, before he was willing to undertake this *Macrocosm* of mine, was eager to know the opinions of many men of letters, both papists and Lutherans as well as Calvinists, and to find out their feelings concerning this volume. Accordingly, it is possible to see this from a portion of this letter:

> Concerning your great volume, before the printer was willing to undertake it, he showed it to many other learned men, who indeed greatly praised your work; furthermore, he even showed it to the Jesuits, who come here in great numbers at bookfair time, and they added their own stamp of approval, with one exception: they felt that the work was most worthy of publication if the topic of Geomancy was deleted, which they, as you know, condemn according to their religion. But we do not consider this judgment of theirs to be the of the least concern, etc. 20th April:

[6ᵛ] In the year of Our Lord 1617
Your most devoted,
Justus Helt[4]

In the same year I received through the hands of a certain foreigner completely unknown to me another letter written by a certain most learned man; I have put forward here to your Royal eyes a portion which pertains to my writings:

By exhibiting to all men a beautiful example of your genius, learning and diligence, you certainly do what is befitting a good man and a philosopher. And certainly it is most reasonable that all good men should hold a high esteem even the ignoble who are adorned with every type of erudition, but particularly those who are distinguished by modesty of mind, temperance, the true religion and piety. There is not much which I could write about you; for it is likely that you already know it well. In distinguishing the harmony of mundane music you profess to be the most practiced and experienced; and this is perhaps not undeservedly, etc.

December 19th 1617.

Most dedicated to you,

Du. Bourdalone[5]

Additionally, I have received a note written by a certain distinguished doctor from Germany, a part of which I have inserted here:

Your writings on the arcane philosophy are very much approved of by their most profound followers. Your works therefore will also constitute additional things which pertain to the Macrocosm and Microcosm, and lead to the very center of things. For undoubtedly you will give an opportunity to other more occult pilosophers to come forward into the public forum, so that at least when these strengths are thus joined together, in this Saturnine age, it will be possible for us, the inquirers, [7ᵛ] to investigate, with open eyes, the heaven and earth as well as all of Nature disrobed of her garment, etc.

Done in Anhalt

31st December 1617

Your most obervant

Mattias Engelhart Philosopher

and Doctor of Medicine

an Anhalt.[6]

From these it is possible for Your Majesty to perceive that even my Physical history has been very well received and accepted by investigators of Chemical Philosophy. This most distinguished Doctor gives excellent testimony to this matter. Also, for the same purpose, I received another letter from a certain Doctor of Law, in

French, this small portion of which I, your most loyal subject, have selected for Your Majesty:

> Please measure my affection by the effects of my devotion, which being aroused primarily by the admiration for your
> [8ʳ] noble mind, advanced not only by the continual consideration of that which is in your treatises and works; the richest part of which, in the manner of the rising sun, now casts its rays on our Germany, as its light does through all the places of the most learned of Europe.
>
> Your most obliged friend and servant
>
> From Vienna 3 February   Jean Balthasar
> in the year 1618   Ursin Bayerius[7]

And although I could offer Your Majesty many similar testimonies selected from the letters of others who wrote much good about me and my works, for the sake of brevity all others have been omitted; only by this note from a more distinguished German and Professor Primarius of the new University of Giessen in Germany shall I impose an end to this *Declaration* of mine to Your Most Serene Majesty and to the defense of my work, the *Macrocosm*. His letter is as follows:

[8ᵛ] Relying on the common bond of the study of Philosophy and Medicine, by which we are linked although otherwise separated by great distances, I have ventured to write this letter to you, most noble and excellent sir, patron much worthy of honor, particularly since an opportunity has been given by the present candidates in medicine, who until now have been my house-guests, etc., and have departed from our University in order to pay their respects to your England and particularly to you. Moreover, as your noble work, the *Macrocosm* (which you have published with the immortalizing of your name), sufficiently and even more gives evidence of singular benevolence and eagerness to help all men of letters to such a degree that I doubt not that you are also quite willing to favor the present learned young men.[8] [9ʳ] According to Cicero, it is a virtue to love that which is invisible, even as now happens to you among all good men. For who does not commend your singular skill in the investigation of things,[9] when you recall from Hell, as it were,

the true principles of things, which Hippocrates of former
days in some places treated unclearly, and Paracelsus of
recent days dealt with superficially; furthermore, you use a
restraint in refuting the opinions of others which is rare
amongst us. Only continue, worthy sir, do not suppress the
description of the Microcosm, whose title promises not vulgar
things; complete the eternal monument to your name such
that it may be accomplished with felicity; and I pray that you
live for a very long time and with good fortune. Written
hurriedly at the new University of Giessen in Hesse.

[9^v] 10 August 1618

Your most devoted
Gregor Horstius Doctor of Medicine
Professor Primarius and
Chief Physician to
Prince Ludwig of Hesse[10]

Certainly from the above letter from this most distinguished man it
is sufficiently obvious that my opinions are not new, but rather are
the most evident explications and most clear demonstrations of the
secrets of nature, which have been concealed or hidden by the
ancient philosophers under the guise of allegorical riddles and
enigmas. On behalf of myself and this book of mine I could mention
many other things filled with the most profound prudence,
knowledge and judgment, which, lest I offer tedium to the ears and
eyes of Your Majesty, I will omit, provided only that I am satisfied
that the brilliance of the writing of these other learned men will
show the measure of their approval for my works corresponding to
their good will toward them, and will make clear to Your Majesty
their opinions [10^r] about my volume, the *Macrocosm*, by this
sincere (as it were) Dedication. Accordingly, O Most Serene King,
I humbly entrust my cause to Your Majesty's distinguished
wisdom and singular kindness, as well as your judgment and just
government, which are filled and adorned with profound knowledge. Your Majesty, who by asserting prominent evidence of your
prudence and justice, certainly informs all men that, although
Astraea (that chaste and virginal goddess of justice), in abandon-
ing the iniquitous earth was borne away on sublime wings, has
ascended into the star-bearing palace of the heavens bearing with
her the scales of justice; nevertheless, she, moved with pity towards

218

us mortal Britons, has left behind here on earth her progeny, descendant, or, as it were, her own Imperial offspring, who will always judge rightly concerning disputed political matters, and will govern with kindness; and in fact in whom the conspicuous gifts of wisdom are so firmly rooted, and so deeply implanted, that, just as he himself blazes with the royal desire to exercise justice, he is provided with Nature's instinct for giving judgments and is thus well prepared in heart to settle whatever is in doubt, inasmuch as his learning and knowledge have rendered him capable of anything. O Most Serene King, endowed with an ineffable grace and majesty, the unexpected splendor of your grace and the serene appearance of Your [10ᵛ] Majesty has moved me, your obedient servant and most unworthy subject, and has excited immense joy in me; therefore I humbly entreat that Your Majesty, with the same grace and great richness of mind, not desist from accomplishing that work most full of divine love, and while I, turning all efforts of my life and studies to the praise of God and honor of Your Majesty, not cease to pour forth prayers to the Creator on behalf of your safety, the security of the Kingdom, and the success of every undertaking. And indeed may there be praise, honor and glory for ever for my King and my Creator, who is both eternal and necessary to support life.

END

NOTES

1 The identity of Justus Helt has not been determined. For the other portion of the letter from Helt, see note 4.
2 Possibly Michael Maier. See Chapter VI.
3 Johann Theodor de Bry and Hieronymous Gallerus in Oppenheim.
4 This portion of Helt's letter was also included in "A Philosophical Key," fols. 12ᵛ–13, with Fludd's English translation as follows:

> As tuching your great Volume, before the printer would undertake it, he shewed it unto many other learned men, which did very much commend your work; Also he made the Jesuits acquainted with it, who in numbers resort unto the fayer of Frakford [sic], which adding also their spur to your commendations sayed, that, only on thing excepted, it was a work most worthy of edition, namely if Geomancy were omitted; the which science (as you know very wel) they mistake of for their religions sake: But we esteeme not of this their latter judgment.

5 The Lord of Bordalone was the Chief Secretary to Charles of Lorraine, fourth Duke of Guise. Fludd styles Bordalone as "amicus meus," and it was probably the noble secretary who arranged for Fludd to tutor the Duke and his brother. See Fludd, *Anatomiae Amphitheatrum* (Frankfurt: de Bry, 1623), p. 233. English translation in John Webster, *A Displaying of Supposed Witchcraft* (London: Jonas Moore, 1677), pp. 319–20.

6 Matthias Englehart was the town physician of Aschersleben, located between Magdeburg and Halle. Aschersleben was the town built adjacent to the ancient seat of the Princes of Anhalt. Fludd includes this letter in the "Philospohical Key," fols. 19–19^v, with this translation:

> Your writings ar very wel approved by the most inward and affectioned serchers of the secret and mystical Philosophy. It wil therefore be your part fully to publish the rest, that belongeth as wel to the great as little world; for in so doing you wil shew other secret and concealed Philosophers the way to appear openly and show themselves to the world, that thereby all their forces being so united, it may at last be permitted unto us (the diligent searchers after Nature) even in this Saturnine age to knowe heaven and earth, and so to uncase and discover that universale nature, which is masked in darkness.

7 The identity of the writer of this letter has not been established.

8 In his translation in the "Philosophical Key," Fludd changes this line to read "thes two learned young men." See note 10.

9 Here Fludd's translation reads "inquisiton of naturall mysterys." See below.

10 Gregor H. Horst (1578–1636) was undoubtedly the most esteemed and famous of those whose letters Fludd incorporates in the "Declaratio Brevis." He received his M.D. from Basel in 1606, and became a Professor of Medicine at Wittenberg. In 1608, the Landgrave of Hesse-Darmstadt called him to the University of Giessen and made Horst his personal physician. Horst gave up his professorship in 1622 to become the town physician in Ulm, where he remained until his death. In his time, Horst enjoyed a great deal of fame. He published numerous books, mostly medical, and his contemporaries hailed him as the "German Aesculapius." August Hirsch, *Bibliographisches Lexikon*, 2nd ed., 5 vols. (Berlin and Vienna: Urban and Schwarzenberg, 1929–1934), 3, p. 304. Fludd also includes Horst's letter in the "Philosophical Key" (fols. 10–11), which reads this way in his translation:

> Worthy sir and my much respected friend (Relying on the common bond of study in Philosophy and Physick, by the which we (being otherwise desjoyned by a great distance of place) are united togeather). I thought it fit at this time to wright unto you especially, occasion being offered by thes present candidates in Physick, which have been heatherto my household guests, etc. who departed from

our University on set purpose to see England, and principally to salute you. And as your notable work of the Great World (which you have published with a perpetual eternishing of your name) hath expressed abundantly your singuler good will and love towards such as ar learned: so I doubt not, but that you will readily favour and assist thes two learned young men. Cicero sayeth, that it is the act of vertue to love thos things we see not, even as now it happeneth to you amongst all good men. For who doth not commend your singular industry in the search and inquisition of naturall mysterys, seeing, that you have revoked as it were out of the pit of hel thos true principles, which Hippocrates long since in some places of his works, and Paracelsus of late dayes have handled but superficially; using that modesty in refuting the opinions of others, which is rare and but seldome found among other men. Go forward then, worthy sir, and suppress not the description of the little world, whose title promiseth things, that are not vulgar, and by that means finish the eternal monument of your renowne which that you may accomplish with felicity live happily.

ROBERT FLUDD'S WILL
(Public Record Office, Prob. II/175)

IN THE NAME OF GOD AMEN the Sixth daie of September in
the yeare of our Lord Jesus Christ one thousand six hundred thirtie
and seaven And in the thirteenth yeare of the raigne of our
soveraigne Lord Charles by the grace of God King of England,
Scotland, France and Ireland, defender of the faith, etc.

I Robert Fludd of the Parish of Saint Katherine Coleman
London Esquire, and Doctor of Physick, being sick of body but of
perfect rememberance, thankes therefore bee given unto Almightie
God, doe ordaigne and make this my last will and testament in
manner and forme as follows:

First I bequeath my Soule and Spirit unto God Almightie the
Author and Creator thereof, noe may doubting of a perfect and
compleat resurrection unto eternall Life, and Salvation by his
omnipotent vertus, and gracious Spirit of my Lord Jesus Christ.

And I leave my body to bee buried in the church belonging to
the Parish Church of Bearsted in the countie of Kent, under that
stone which is alreadie layed for me in that church.

And my desire is that a new superscription bee put on my Grave-
stone, the old being taken off, and that there bee a Monument
sett up for me in that church neare unto the forme of M. Camden's
Monument in Westminster with seaven distinct volumes. Upon the
first of which shalbe written, PHISICA ET TECHNICA MACRO-
COSMI HISTORIA, upon the second PHISICA ET TECHNICA
MICROCOSMI HISTORIA, upon the third MISTERIUM
CABALISTICUM,[1] upon the fowerth, AMPHITHEATRUM
ANATOMICUM, upon the fift, PHILOSOPHIA SACRA, upon
the sixt MISTERIUM SANITATIS,[2] and upon the seaventh,
MISTERIUM MORBORUM.[3]

And withall I doe request, yea, and I doe charge mine executor
hereafter named not to suffer my body to bee violated or rutt open
but that it bee onely wrapped in cereclothes, and afterwards put
according to custome in a new winding sheete freshly brought for
that purpose, and soe to bee laid in a coffin stuffed with Bran, and
Lime, etc.

And as for the Scutchings and other necessary ceremonies which shall be done by the advise of a harrold att arms, I leave that also, and all other things also belonging thereunto to be charged with creditt unto mine executors. Also my wish is that all duties belonging unto the Parish Church of Saint Katherine Coleman aforsaid be discharged before my body or corps doe depart from my house.

And I doe give, and bequeath unto the poore of the said Parrish, three pounds of lawfull money of England, there to be distributed among them the next sundaie after my bodie is carryed away.

And my desire is that as soone as may bee my body or corps be carreyed downe into the countrey by coach, and that in the night time if it may be and that soe many coaches bee hired as wilbe sufficient to carry all such as I shall give Mouning unto, to Accompany my corps unto the Grave, and that all of them be decently entertained upon the way at my cost & expense.

Also my desire is that fifty torches or more, as shall appear fitt to my executors and overseers shall attend my corps from the Inne at Bearested unto the church, and soe bee interred in the Nighttime, if it bee possible, but if that bee prohibited by the harrold, then to bee buried in the afternoone, and that without anie further ceremonies onely then wine and Beefitting stuffe being bestowed on such as are present.

Item. I give and bequeath unto the poore of the Parrish of Bearsted the somme of three pounds of lawfull money of England.

And whereas a good part of my temporal estate, and lyvinge is but during my natural life, being issueing in forme, and manner of an yearely Annuity partly out of my late father Sir Thomas Fludd his Lande, and partlie out of the Exchequer of one Mr. Paule Portis,[4] and they doe cease aft my death.

Now my wish and pleasure is that all my estate aswell in fee simple as Leases, money, moveables and household stuff whatsoever shalbe bequeathed, disposed of, and distributed as followeth.

And I doe make, constitute, nominate and appoint my Nephew Thomas Fludd[5] Esquire, and my kinsman Arthur Trevor Esquire[6] lawfull executors of this my last will and Testament.

Item. I give, and bequeath to my said kinsman Arthur Trevor the somme of fiftie pounds of lawfull money of England, to bee paid unto him twelve moneths next after my Decease.

Item. I give unto him my two houses.

223

Item. I give and bequeath unto him, one cloake of unshorne velvet with fower broad Imbrodered Laces, and Lyned with plush which is perfectly new and not worne, and cloath for a mourninge suite, and cloake.

Item. I give and bequeath unto my aforesaid Nephew Thomas Fludd all my estate which I have in my Dwelling house in the Paris of St. Katherine Coleman aforesaid which cost mee fower hundred pounds besides one hundred pounds in building and repayringe, and is now well worth threescore pounds a yeare, paying but a pepper corne a yeare.

Item. I give and bequeath unto my said Nephew Thomas Fludd all my Lande in the county of Suffolke called Oxlande with all Barnes, profitts and comodityes thereunto belonging, to have and to hold unto him and his heires forever.[7]

Item. I give and bequeath unto his Brother Lywing Fludd,[8] All my Studdie of Bookes, being more fitt for his use than anie others.

Item. I give, and bequeath unto the said Lyving Fludd Twenty Pounds of lawfull money of England, to be paid unto him within twelve months next after my Decease.

Item. I give and bequeath unto my Nephew and Godsonne Robert Fludd[9] all the estate which I have in anie house and land in Allington, and Romney Marshe, and all my estate which I have in anie land in Stanford in the Countie of Kent, with my estate in the land lying by Weston Hanger Porte pale with all my estate in the upper Wood Land now being in the occupation of one Coleman.[10]

Item. I give and bequeath unto my said Nephew and Godsonne Robert Fludd the somme of Fiftie Pounds of lawful money of England to bee paid unto him within twelve months next after my Decease. But if the said Robert Fludd shall die or depart this mortall life before he shall returne againe into England, then I doe hereby will and bequeath all such estate which I bequeathed unto him in house and land as aforesaid unto his Brother Living Lloyd,[11] and the Fifty Pounds to remaine to my said Executor Thomas Fludd.

And my will and desire is that their Mother shall have, receive and take to her owne use, behouse the benefitts and profitts which shall arise of the said house, and land from the time of my decease, until the said Robert shall returne to England, or that there be a sufficient testimonie that the said Robert Fludd is dead.

Item. I give and bequeath unto the Colledge of Phisitians the some of Twenty Pounds of lawfull money of England, to be paid within three months next after my decease.

Item. I give and bequeath unto my Anncient servant Thomas Millett all the estate which I have in anie house or land in Weavering in the County of Kent, with all benefitts, and profitts of the Fulling Earth Pitts, as it is now in the occupation of one _____ Thad of Maidstone.[12]

Item. I give, and bequeath unto the said Thomas Millett all my estate which I have in and land at Doting Hill, now in the occupation of William Hartridge.[13]

Item. I give and bequeath unto the said Thomas Millett all such debts as shalbee owing unto me att the time of my death for Physicke.

Item. I give and bequeath unto the said Thomas Millett a Mourning Suite, and Cloake if that he accompany my Corps unto the Buryall.

Item. I give, and bequeath unto Anne the wife of the said Thomas Millett the somme of Twenty Pounds of lawfull money of England, to be paid unto her within twelve months next after my decease.

Item. I give, and bequeath unto her good plaine mourninge, soe that she accompany my Corps unto the Buryall.

Item. I give, and bequeath unto my sister Katherine Lunsford[14] the somme of Twenty Pounds of lawfull money of England to buy her a peece of Plate, to be paid unto her within twelve months next after my Decease.

Item. I give and bequeath unto her Stuffe fitting for her mourninge.

Item. I give and bequeath unto my Brother Mr Thomas Lunsford cloth for a mourninge Suite and cloake.

Item. I give and bequeath unto my Sister Sarah Broomefeilde the somme of Twenty Pounds of lawfull money of England to buy her a peece of plate, to bee paid unto her within twelve months next after my decease.[15]

Item. I give and bequeath unto her Stuffe fitting for her mourning.

Item. I give, and bequeath unto the Lady Elizabeth Mustian[16] Tenn Pounds of lawfull money of England to buy her a peece of plate.

Item. I give and bequeath unto her Stuffe fitt for her mourninge, soe that she accompanie my Corps unto the Buryall.

Item. I give, and bequeath unto Doctor Denton[17] cloth for a mourninge cloake and suite, soe that he goe with my Corps to the Buryall.

Item. I give, and bequeath unto the Widdowe of my brother John Fludd deceased the somme of Tenn Pounds of lawfull money of England, to bee paid unto her twelve months after my decease.[18]

Item. I give and bequeath unto her good plaine Mourning soe that she accompany my Corps unto the Buryall.

Item. I give and bequeath unto her daughter the some of Five Pounds of lawfull money of England, to be paid unto her the life space of twelve months next after my Decease.

Item. I give and bequeath unto my cousin Katherine Ward[19] the somme of Five Pounds of lawfull money of England, to bee paid unto her twelve months next after my Decease, to buy her a peece of plate.

Item. I give unto her the best of my three Rings which I have, and my will and desire is that she shall take her choise of them all.

Item. I give and bequeath unto her Stuff fitt for her mourning, soe that she accompanie my Corps unto the Buryall.

Item. I give and bequeath unto my Cousin Sarah Kerick the some of Five Pounds of lawfull money of England to buy her a peece of plate.[20]

Item. I give unto her the next best of my three Rings, and my will, and my desire is that she shall take her choise of them.

Item. I give and bequeath unto her Stuffe fitt for her mourninge, soe that she accompanie my Corps unto the Buryall.

Item. I give and bequeath unto my Goddaughter Sarah Kerrick the somme of Tenn Pounds of lawfull money of England, to bee paid unto her twelve months after my Decease.

Item. I give and bequeath unto her my other Ring.

Item. I give and bequeath unto her Stuff fitt for her mourning soe that she accompany my Corps unto the Buryall.

Item. I give and bequeath unto my servant Walter Powell Tenn Pounds of lawfull money of England, to be paid unto him presentlie after my Decease.

Item. I give and bequeath unto my said servant Walter Powell cloth for a mourning cloake and suite.

Item. I give and bequeath unto my Cousin Katherine Vaughn[21]

Stuff fitt for her mourninge, soe that she accompany my Corps unto the buryall.

Item. I give and bequeath unto my Cousin Mrs. Mary Fuller[22] the somme of Five Pounds of lawful money of England to buy her a Ringe, to bee paid unto her one yeare after my Decease.

Item. I give and bequeath unto her Stuffe fitt for her mourninge, soe that she accompany my Corps unto the Buryall.

And I doe nominate and appointe my Cousin Ward Draper,[23] and one of his Masters receivors, my Cousin Kenneth Silkeman[24] and my aforesaid anncient servant Thomas Millett Overseers of this my last will and testament, intreating them to be ayding and assisting unto my Executors before named in due performance of this my last will and testament.

And I doe hereby give and bequeath unto either of my said Cousins cloth for a mourning suite, and cloake.

And my will and desire is that my said Cousin Warde shall saive all the cloth which shalbe given by this my last will for Mens mourninge, and my said Cousin Kerrick all that shalbe given for womens mourninge.

All the rest and residue of my land, tenemente, goods, chattells, plate, readie money, and household stuff whatsoever not before bequeathed, my debte being first paid, and my funerall expenses discharged, I give and bequeath to the aforesaid Thomas Fludd, and of my Executors before named, charging him, and my other Executor Arthur Trevor, that they doe in all things justly, and duely performe this my last will in all points, and that they doe presently after my Decease cause all those to be made knowne thereof unto whome I have hereby given mourning that they may the better prepare themselves to goe to my buriall, if they intend soe to doe.

Lastly I doe renounce, and herby revoke all former wills by me made and doe declare this to be my last will and testament containing three sheets of paper and this parte of a sheete, to everyone of which I have hereunto putt my hand and seale the day and yeare herein written.

Robert Fludd

Signed, Sealed published and declared to be the last will and testament of the said Robert Fludd, Esquire, Doctor of Phisick, the daye and yeare first thereon written in the presence of Edward Lyvington, and of me Jo. Butler Scr.

Item. I further will and bequeath unto M. Lamdon[25] the some of Fortie Shillings of lawfull money of England. This was acknowledged to bee parte of the will in the presence of Edward Livington and of me Jo. Butler Scr.

NOTES

1 Book II, Tractate II of the Microcosmical History: *Tomi secundi tractatus secundus, de praeternaturali utriusque mundi historia* (Frankfurt: Johann Theodore de Bry, 1621).

2 *Medicina Catholica*, Book I, Tractate I (Frankfurt: William Fitzer, 1629).

3 *Medicina Catholica*, Book I, Tractate II, bearing the title *Integrum Morborum Mysterium* (Frankfurt: William Fitzer, 1631). It is interesting to note how Fludd divides his own works into seven parts.

4 He received several annuities from the will of his father, Sir Thomas Fludd (Public Record Office, Prob. II/110, dated 18 February 1606).

5 Thomas Fludd (1605–16?) was the son of Thomas Fludd (d.1612), Robert's older brother, who succeeded to the family seat, Milgate House, on the death of Sir Thomas in 1607. Robert's nephew moved the family seat to Gore Court in nearby Otham (which still stands) and sold Milgate House in 1624. The nephew Thomas was Sheriff of Kent from 4 November 1651 to 12 November 1652 ("List of Sheriffs for England and Wales," *Lists and Indexes*, No. IX, p. 70).

6 Not otherwise identified.

7 This is presumably the grant given to him by Charles I on 8 June 1629 (see Chapter IV, note 16). I have been unable to locate this estate.

8 Lywing, also spelled Lyving, Lewin, Lewen, Levin and Livin, Fludd (c.1612–1678), younger brother of Thomas (above, note 5), received his B.A. from Trinity College, Cambridge 1631/2, and his M.D. from Padua 25 August 1639 (John and J. A. Venn, *Alumni Cantabrigiensis*, I, 153). He practiced medicine in Loose and Maidstone in Kent, and was incorporated M.D. at Oxford 21 November 1661. Levin Fludd is buried at the parish church at Loose (Loose Parish Records, Cathedral Archives and Library, Canterbury). Since Levin received his uncle's library and was a graduate of Trinity, it is possible that he donated the "Philosophical Key" MS. to his alma mater. The administration record of Levin's estate is sparse and does not give a detailed dispositon of his goods. (Act Books of the Consistory Court of Canterbury, entry dated 4 November 1678; Cat. Mk. PRC 22/20/63, Kent Archives Office, Maidstone).

9 A younger brother of Thomas and Levin Fludd, not otherwise identified.

10 Land received from Sir Thomas Fludd, note 4.

11 Levin Fludd, note 8 above.

12 Also descended from Sir Thomas. First name of Thad blank in will.

13 Also part of Sir Thomas' estate.

14 Younger sister of Robert (b. 1579), married Thomas Lunsford, Esq., 7 March 1603.

15 Another younger sister (b. 1581), first married to Henry Buffkin, Esq., 19 January 1601, the same day as her brother Thomas' wedding to Katherine Buffkin, Henry's sister. Their father was Levin Buffkin of Otham, a member of Gray's Inn, as was son Henry. Sarah later married a Broomefield.

16 Not otherwise identified. Perhaps a member of Fludd's household.

17 Not otherwise identified.

18 John was an older brother (b. 1572), and was B.A. St. John's Oxford, 1593/4. His wife's name was Marie.

19 Not otherwise identified.

20 Ditto.

21 Ditto.

22 Ditto.

23 Ditto.

24 Ditto.

25 Ditto.

APPENDIX C

A transcription of the list of books given by Robert Fludd to Jesus College, Oxford, as listed in the Benefactors' Book.

ANNO 1630

Robert fludd alias de fluctibus D^r. in phisicke of the universitie of Oxon.: and descended out of Wales gave these books following:

1	Concordantiae Bibliorum Lat.	4°
2	Rathborne Survayor.	fo.
3	Bungus. Latin	4°
4	Carbo Latin	8°
5	Alumo Lat.	fo.
6	Palladius Lat.	fo.
7	Pierij Hieroglyph.	fo.
8	Keperlus: Lat.	fo.
9	Chroniq de france vol: 3^{bus}	fo.
10	Terentius Galice	fo.
11	Augustin' de civitate dei gall.	fo.
12	Biblia Castalion.	fo.
13	L. Anatomie	fo.
14	Aristot: Rhet: Goulston	4°
15	Theatrum Chemitû vol. 4^{or}	8°
16	Mercatoris Geograph: Lat.	fo.
17	Dordoneus de stirpibus Lat.	fo.
18	Shonerus: Italien	4°
19	Theatrū philos: christianū:	4°
20	Copernicus de revol: orbiū coel.	fo.
21	Biblia Pagnini & Vatabli.	fo.
22	Antonius Maria mediciñ	fo.
23	Mal: Rempis	8°
24	J. David Rhesj camb: ling: institut.	fo.

Of the above books, the following were tentatively identified as still in the College's collection:

1 Either Antwerp 1585 (Plantin), or Hanover 1618.

2 Aaron Rathborne, *The Surveyor* (London, 1616).

3 Petri Bungi (or Bongi), *Numerorum Mysteria* (Paris, 1617).

5 ? Erroneously, for P. Alunno, *Della fabrica del mundo* (Venice, 1562)

6 Andrea Palladio, *I quattro libri dell' architettura* (Venice, 1616)

7 John Pierius Valerianus, (Basel, 1567); or *Commentaires hiero-glyphiques* (Lyons, 1576).

8 *Prodromus dissertationum cosmographicarum* (Frankfurt, 1621).

9 *Chroniques de France.* 3 vols. (Paris, 1493).

10 Paris, ?1500.

11 Paris, 1585 (trans. G. Hervet).

12 Basel, 1554.

14 London, 1619.

15 *Theatrum Chemicum* (ed. L. Zetzner), 4 vols. (Strasbourg, 1613).

16 Duysburg, 1578.

18 Joannes Schonerus, *I tre Libri della nativita* (Vinegia, 1554).

APPENDIX D

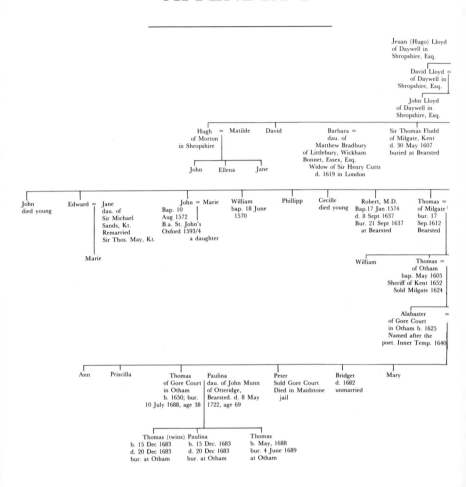

Jeuan (Hugo) Lloyd
of Daywell in
Shropshire, Esq.

David Lloyd =
of Daywell in
Shropshire, Esq.

John Lloyd
of Daywell in
Shropshire, Esq.

Hugh = Matilde   David
of Morton
in Shropshire

John   Ellena   Jane

Barbara =
dau. of
Matthew Bradbury
of Littlebury, Wickham
Bonnet, Essex, Esq.
Widow of Sir Henry Cutts
d. 1619 in London

Sir Thomas Fludd
of Milgate, Kent
d. 30 May 1607
buried at Bearsted

John
died young

Edward = Jane
dau. of
Sir Michael
Sands, Kt.
Remarried
Sir Thos. May, Kt.

Marie

John = Marie
Bap. 10
Aug 1572
B.a. St. John's
Oxford 1593/4
a daughter

William
bap. 18 June
1570

Phillipp

Cecille
died young

Robert, M.D.
Bap.17 Jan 1574
d. 8 Sept 1637
Bur. 21 Sept 1637
at Bearsted

Thomas =
of Milgate
bur. 17
Sep.1612
Bearsted

William

Thomas =
of Otham
bap. May 1605
Sheriff of Kent 1652
Sold Milgate 1624

Alabaster   =
of Gore Court
in Otham b. 1625
Named after the
poet. Inner Temp. 1640

Ann   Priscilla

Thomas
of Gore Court
in Otham
b. 1650; bur.
10 July 1688, age 38

Paulina
dau. of John Munn
of Otteridge,
Bearsted. d. 8 May
1722, age 69

Peter
Sold Gore Court
Died in Maidstone
jail

Bridget
d. 1682
unmarried

Mary

Thomas (twins) Paulina
b. 15 Dec 1683
d. 20 Dec 1683
bur. at Otham

b. 15 Dec. 1683
d. 20 Dec 1683
bur. at Otham

Thomas
b. May, 1688
bur. 4 June 1689
at Otham

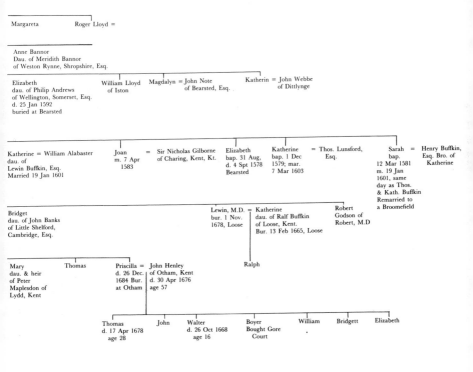

Margareta   Roger Lloyd =

Anne Bannor
Dau. of Meridith Bannor
of Weston Rynne, Shropshire, Esq.

Elizabeth   William Lloyd   Magdalyn = John Note   Katherin = John Webbe
dau. of Philip Andrews   of Iston   of Bearsted, Esq.   of Dittlynge
of Wellington, Somerset, Esq.
d. 25 Jan 1592
buried at Bearsted

Katherine = William Alabaster   Joan   = Sir Nicholas Gilborne   Elizabeth   Katherine   = Thos. Lunsford,   Sarah   = Henry Buffkin,
dau. of   m. 7 Apr   of Charing, Kent, Kt.   bap. 31 Aug,   bap. 1 Dec   Esq.   bap.   Esq. Bro. of
Lewin Buffkin, Esq.   1583   d. 4 Spt 1578 1579; mar.   12 Mar 1581   Katherine
Married 19 Jan 1601   Bearsted   7 Mar 1603   m. 19 Jan
                                                                                                                               1601, same
                                                                                                                               day as Thos.
                                                                                                                               & Kath. Buffkin
                                                                                                                               Remarried to
                                                             Lewin, M.D. = Katherine   Robert   a Broomefield
Bridget   bur. 1 Nov.   dau. of Ralf Buffkin   Godson of
dau. of John Banks   1678, Loose   of Loose, Kent.   Robert, M.D
of Little Shelford,   Bur. 13 Feb 1665, Loose
Cambridge, Esq.

Mary   Thomas   Priscilla = John Henley   Ralph
dau. & heir   d. 26 Dec. of Otham, Kent
of Peter   1684 Bur.   d. 30 Apr 1676
Maplesdon of   at Otham   age 57
Lydd, Kent

                             Thomas   John   Walter   Boyer   William   Bridgett   Elizabeth
                             d. 17 Apr 1678   d. 26 Oct 1668 Bought Gore
                             age 28   age 16   Court

BIBLIOGRAPHY

I. OUTLINE OF THE *UTRIUSQUE COSMI . . . HISTORIA*

Volume I
The History of
the Macrocosm

Tractate I (UCH I, 1)
Utriusque cosmi . . . historia.
(Oppenheim: de Bry, 1617)

Tractate II (UCH I, 2)
Tractatus Secondus, De Naturae Simia
(Oppenheim: de Bry, 1618)

Volume II
The History of
the Microcosm

Tractate I

Section I (UCH II, 1, i)
Tomus Secundus De Supernaturali . . .
(Oppenheim: de Bry, 1619)

Section II (UCH II, 1, ii)
Tomi Secundi Tractatus Primi . . .
(Oppenheim: de Bry, 1620?)

Tractate II

Section I

Portions I & II (UCH II, 2, i,1 & 2)
Tomi Secundi . . . De Praeternaturali . . .
(Frankfurt: de Bry, 1621)

Portion III (UCH II,2,i,3)
Anatomiae Amphitheatrum
Frankfurt: de Bry, 1623

Portion IV (UCH Ii,2,i,4)
Philosophia sacra
Frankfurt: Officina Bryana, 1626

Sections II & III (not published)

Tractate III (not published)

II. OUTLINE OF THE *MEDICINA CATHOLICA*

Volume I

Tractate I (MC I, 1)
Medicina Catholica, Sanitatis Mysterium
(Frankfurt: William Fitzer, 1629)

Tractate II

Section I (MC I, 2, i)
Integrum Morborum Mysterium
(Frankfurt: William Fitzer, 1631)

Section II (MC I, 2, ii)
KATHOLIKON MEDICORUM KATOPTRON
(Frankfurt: William Fitzer, 1631)

Section II, Portion III, Part III
Pulsus   (MC I, 2, ii, 3, 3)
(Frankfurt: William Fitzer, 1631?)

Volume II
Medicamentosum Apollinis Oraculum
(Outline published at end of Volume I, Tractate II, dated 1630, but Volume II itself not published)

III. PRIMARY SOURCES

A. Manuscripts

Cambridge. Trinity College Library. Western MS. 1150. Fludd, "A Philosophical Key." Published edition: Allen G. Debus, *Robert Fludd and his Philosophical Key*. New York: Science History Publications, 1979.

London. British Library. Royal MS. 12 C ii. Fludd, "Declaratio Brevis." Published edition: Translation by William H. Huffman and Robert A. Sellinger, Jr, introduction by William H. Huffman, "Robert Fludd's 'Declaratio Brevis' to James I," *Ambix* 25 (1978), 69–92.

London. British Library. Sloan MS. 2149. Baldwin Hamey the Younger, "Bustorum aliquot Reliquae."

New Haven. Yale University Library. Miscellaneous Manuscript 170. Filmer MS. 3. Musical compositions by Fludd. See Todd Barton, "Robert Fludd's Temple of Music: A Description and Commentary," M.A. Thesis, University of Oregon, 1978; Appendix III, 202–10.

Oxford. Bodleian Library. MS. Ashmole 766 and MS. Ashmole 1507. Fludd, "Truth's Golden Harrow." Published edition: C. H. Josten, "Truth's Golden Harrow," *Ambix* 3 (1948), 91–150.

B. Fludd's Printed Works

Fludd, Robert. *Anatomiae amphitheatrum effigie triplici, more et conditione varia designatum*. Francofurti: Sumtibus Johannis Theodori de Bry, 1623. (UCH II,2,i,3) This is a continuation of the *History of the Microcosm*; see below and outline of *Utriusque cosmi . . . historia*.

——. *Apologia compendiaria. Fraternitatem de Rosea Cruce suspicionis et infamiae maculis aspersam, veritas quasi Fluctibus abluens et abstergens*. Leyden: Gottfridum Basson, 1616.

——. *Clavis Philosophiae et Alchemiae Fluddanae*. Francofurti: Apud Guilhelmum Fitzerum, 1633.

——. *Discursus de unguento armario*. In Rattray, Sylvestri, ed. *Theatrum Sympatheticum Auctum, exhibens varios authores*. Norimbergae: Apud Johan. Andreae Endterum & Wolfgangi Junioris Haeredes, 1622.

——. *Doctor Fludds Answer unto M. Foster, or the squeesing of Parson Fosters Sponge, ordained by him for the wiping away of the Weapon-Salve*. London: Nathanel Butter, 1631.

———. *Responsum ad Hoplocrisma-spongum M. Fosteri . . .* Goudae: Petrus Rammazenius, 1638. (Latin translation of above.)

———. *Medicina Catholica, seu mysticum artis medicandi sacrum. In tomos divisum duos.* (M.C. I,1) Francofurti: Typis Caspari Rotellii, Impensis Wilhelmi Fitzeri, 1629. (This edition contains the Tractatus Primus of the Tomus Primus bearing the title "Sanitatis Mysterium." The remainder of the *Medicina Catholica* is under the title *Integrum Morborum Mysterium.*)

———. *Integrum Morborum Mysterium: sive Medicinae Catholicae tomi primi tractatus secundus in Sectiones distribus duas.* (M.S. I, 2, i) Francofurti: Typis excusis Wolfgangi Hofmanni, Prostat in Officina Guleilmi Fitzeri, 1631. (This continuation of the *Medicina Catholica* contains three treatises belonging to the Tomus Primus and a fold-out schematic of the Tomus Secundus, which treatise is the Sectio Prima of the Tractatus Secundus, "Integrum morborum, seu meteororum insalubrium mysterium." Sectio Secundus is "KATHOLIKON MEDICORUM KATOPTRON, sive tomi primi, tractatus secundi, sectio secunda, De Morborum Signis," and bears the date 1631 on the title page. (M.C. I, 2, ii) The third treatise, with a separate title page, has the title "PULSUS, seu nova et arcana Pulsuum historia, esacro fonte radicaliter extracta, nec non medicorum ethnicorum dictis & authoritate comprobata, hoc est, portionis tertiae pars tertia of the Sectio Secundus De pulsuum Scientia," n.p., n.d. [p. 93 bears the date 1629]. (M.C. I, 2, ii, 3, 3) The Section Tertius was never printed. The fold-out schematic for the Tomus Secundus is bound at the end with the title "Medicamentorum Apollinis Oraculum, hoc est, Medicinae Catholicae, seu mysticae medicandi artis, Tomus Secundus," n.p.: Typus excudebatur Wolfgangi Hofmanni, 1630. This entire volume was reissued under the title *Meteorum insalubrium mysterium.* Moguntiae [Mainz]: L. Bourgeat, 1682.

———. *Monochordum Mundi Symphoniacum, seu, Replicatio Roberti Flud . . . ad Apologiam . . . Johannis Kepleri . . .* Francofurti: Typis Erasmi Kempferi, Sumptibus Ioan. Theodor. de Bry, 1623.

———. *Philosophia Moysaica.* Goudae: Petrus Rammazenius, 1638.

———. *Mosaicall Philosophy.* London: Humphrey Moseley, 1659. (English translation of above. Internal evidence [p. 232] suggests Fludd wrote most of this work in 1630 or 1631. On p. 287 he mentions visiting a patient in 1637, so the work may have been finished only shortly before his death.) Reprint edition of Books One and Two of the Second Section: Adam McLean, ed., Edinburgh: Magnum Opus Hermetic Sourceworks No. 2, 1979.

———. *Philosophia sacra et vere Christiana seu Meteorologia Cosmica.* Francofurti: Prostat in Officina Bryana, 1626. (UCH I,2,i,4) This volume is also a continuation of the *History of the Microcosm;* see below and outline of the *Utriusque cosmi . . . historia.*

———. *Sophie cum moria certamen, in quo, lapis lydius à falso structore, Fr. Marino Mersenno, Monacho, reprobatus, celeberrima Volumnis sui Babylonici (in Genesi) figmenta accurate examinat.* n.p., 1629. (Bound with the *Philosophia Sacra* above, with separate pagination.)

———. *Tractatus apologeticus integritatem Societatis de Rosea Cruce defendens.*

Ludguni Batavorum [Leyden]: Godefridum Basson, 1617. A German translation of this work is by Adam Michael Birkholz [Ada Mah Booz, pseudonym]. *Schutzschrift für die Aechtheit der Rosenkreutzergesellschaft.* Leipzig: A. E. Boehme, 1782.

———. *Tractatus theologo-philosophicus, In Libros tres distributus, Quorum I de Vita, II de Morte, III, de Resurrectione . . . à Rudolfo Otreb Britanno.* Oppenheimii: Typis Hieronymi Galleri, Impensis Joh. Theod. de Bry, 1617.

———. *Utriusque cosmi maioris scilicet et minoris metaphysica, physica atque technica historia . . . Tomus primus, De Macrocosmi Historia in duos tractatus diuisa.* Oppenheimii: Aere Johan-Theodori de Bry, Typis Hieronymi Galleri, 1617. (UCH I, 1) The first issue of the Technical, Metaphysical and Physical History of the Macrocosm and Microcosm contains the Tractate One of Volume One (the *History of the Macrocosm*), "De mataphysico macrocosmi et creaturarum illus ortu et de physico macrocosmi in generatione et corruptione progressu." English trans-lation of Books One and Two: Fludd. *The Origin and Structure of the Cosmos.* tr. Patricia Tahil. Edinburgh: Magnum Opus Hermetic Sourceworks No. 13, 1982.

———. *Tractatus Secundus, De Naturae Simia seu Technica macrocosmi historia, in partes undecim divisa.* In Mobili Oppenheimo: Aere Iohan-Theodori de Bry, Typis Hieronymi Galleri, 1618. (UCH I, 2) This second tractate of the *History of the Macrocosm* contains eleven essays on man's arts, through the use of which he becomes the "ape of nature." They are: I. De arithmetica universali. II. De templo musicae. III. De geometrica. IV. De optica scientia. V. De arte pictoria. VI. De arte militare. VII. De motu. VII. De tempore. IX. De cosmograpia. X. De astrologia. XI. De geomantia. Reissued, Francofurti: Sumptibus haeredum Johannis Theodori de Bry, Typis Caspari Rôtelli, 1624. The section on geomancy appeared under the title, "De animae intellectualis scientia, seu, Geomantia hominibus appropriata," in *Fasciculus Geomanticus, in quo varia variorum opera geomantica continentur.* Veronae, 1687. 2nd edition, Veronae, 1704. The same section was translated into French by Pierre Vincent Piobb, *Traité de Géomancie (De Geomantia).* Paris: Dangles, 1947. Piobb also translated the section on astrology: *Étude du macrocosme, annotée et traduite pour la première fois par Pierre Piobb, Traité d'astrologie générale (De astrologia).* Paris: H. Daragon, 1907.

———. *Tomus Secundus de supernaturali, naturali, praeternaturali et contranaturali microcosmi historia, in tractatus tres distributa.* Oppenheimii: Impensis Iohannis Theodori de Bry, Typis Hieronymi Galleri, 1619. (UCH II, 1, i) This is Section One of Tractate One of Volume II (the *History of the Microcosm*). It is divided into thirteen books dealing with divine harmonics as they relate to man, the microcosm. At the end of Book IX there is an errata sheet and index. Immediately following is a title page and new pagination for the second section of the first treatise, bearing the title *Tomi Secundi, tractatus primi, sectio secunda, de technica Microcosmi historia, in Portiones VII divisa,* n.p., n.d. (UCH II,2,i) Section II contains essays on Prophesy, Geomancy, The Art of Memory, Astrology, Physiognomy, Chiromancy and the Pyramid. [Beginning with p. 204 and continuing the new pagination, Books X–XIII Of the Sectio Prima appear, ending on p. 277.]

——. *Tomi secundi tractatus secundus, de praeternaturali utriusque mundi historia, In sectiones tres divisa* . . . Francofurti: Typis Erasmi Kempfferi, Sumptibus Joan. Theodori de Bry, 1621. (UCH II,2,i,1 & 2) This continuation of the Microcosm History contains Portions I and II of Section I; Portions III and IV appeared under separate titles as *Anatomiae Amphitheatrum* (UCH II,2,i,3) and *Philosophia Sacra* (UCH II,2,i,4). Sections II and III never appeared.

——. *Veritatis proscenium* . . . *seu demonstratio quaedam analytica, in qua cuilibet comparationis particulae, in appendice quaedam J. Kepplero, nuper in fine Harmoniae suae Mundanae edita* . . . Francofurti: Typis Erasmi Kempfferi, Sumtibus Joan. Theodor. de Bry, 1621.

C. Selected writings of Fludd's Contemporaries

Ashmole, Elias. *Theatrum Chemicum Britannicum*. London: J. Grismond for Nathaniel Brooke, 1652. Reprint, New York: Johnson Reprint, 1967.

Bacon, Francis. *The Philosophical Works of Francis Bacon*. Edited by John M. Robertson. London: Routledge, 1905.

Boissard, Jean-Jacques. *Bibliotheca sive thesaurus virtutis et gloriae*. Frankfurt: William Fitzer, 1628–30.

Bruno, Giordano. *La Cena de la cerneri. The Ash Wednesday Supper*. Edited and translated by E. A. Gosselin and L. S. Lerner. New York: Mouton, 1977.

Culpepper, Nicholas. *Mr. Culpepper's Ghost*. London: Peter Cole, 1656.

Dee, John. *A True and Faithful relation of what passed for many years between Dr. John Dee and some spirits*. London: T. Garthwait, 1659.

——. *John Dee on Astronomy, Propaedeumata Aphoristica*. Edited and translated by Wayne Shumaker. Berkeley: University of California Press, 1978.

——. *The Mathematical Praeface to . . . Euclid . . . (1570)*. Edited by Allen Debus. New York: Science History Publications, 1975.

Dechesne, Joseph. *The Practise of Chymicall, and Hermeticall Physicke for the preservation of Health*. Translated by Thomas Tymme. London: Thomas Creede, 1605.

Durelle, P. Jean. *Effigies contracta Roberti Fludd Angeli* . . . Paris: Apud Guillelmum Baudry, 1636.

Foster, William. *Hoplocrisma Spongus, or a Sponge to wipe away the weapon-salve*. London: Thomas Cotes, 1631.

Frizius, Joachim. *Summum Bonum*. n.p. (Frankfurt): 1629. A German translation of Book IV is in F. Freudenberg, *Paracelsus und Fludd*. Berlin: Hermann Barsdorf Verlag, 1918, 233–71.

Gaffarel, James. *Unheard-of-Curiosities*. London: G.D. for Humphrey Moseley, 1650.

Gassendi, Pierre. *Epistolica exercitatio in qua principia philosophiae Roberti Fluddi, medici, reteguntur, et ad recentes illius libros adversus R.P.F. Marinum*

Mersennum . . . respondetur. Paris: S. Cramoisy, 1630.

——. *Examen philosophiae Roberti Fluddi.* Opera Omnia. 6 vols. Florence, 1727.

Gilbert, William. *De magnete.* London: P. Short, 1600.

Hakewill, George. *An Apologie or Declaration of the Power and Providence of God in the Government of the World.* 3rd ed. London: Robert Allott, 1635.

Hart, James. *KLINIKH, or the Diet of the diseased.* London: John Beale, 1633.

Harvey, William. *De motu locali animalium.* Edited and translated by Gwenneth Whitteridge. Cambridge: The University press, 1959.

Kepler, Johannes. *Adversus demonstrationem . . . de Fluctibus.* In *Gesammelte Werke.* 18 Vols. Munich: C. H. Beck, 1937–1949.

——. *Harmonices mundi.* In *Gesammelte Werke* above.

Lambye, John Baptista. *A Revelation of the Secret Spirit Declaring the Most concealed secret of Alchemie.* Translated by R.N.E. London: Iohn Haviland of Henrie Skelton, 1629.

Lanovius (François de la Noue). *Ad Reverendum Patrem Marinum Mersennum Francisci Lanovii Judicium de Roberto Fluddo.* Paris, 1630.

Mersenne, Marin. *Correspondence du P. Marin Mersenne.* Paris: Presses Universitaires de France, et al, 1945– .

——. *La verité des sciences.* Paris: Toussaint Du Bray, 1625. Reprint, Stuttgart-Bad Canstatt: Frommann, 1969.

——. *Quaestiones celeberrimae in Genesim.* Paris, 1623.

Pharmacopoeia Londinensis. London: E. Griffin, 1618.

A Physical Directory, or a translation of the London Dispensatory made by the College of Physicians in London. London: Peter Cole, 1649.

Reuchlin, Johannes. *De arte cabalistica.* Stuttgart: Frommann Verlag, 1964.

Ridley, Mark. *A Short Treatise of Magneticall Bodies and Motions.* London: Nicolas Okes, 1613.

Scot, Patrick. *The Tillage of Light . . .* London: William Lee, 1623.

Selden, John. *Titles of Honor.* London: W. Stansby for I. Helme, 1614.

Sennert, Daniel. *The Weapon-Salves Maladie.* London: John Clark, 1637.

Thornborough, John. *ΛΙΘΟΘΕΩΡΙΚΟΣ; sive Nihil, aliquid, omnia, antiquorum sapientium vivis coloribus depicta . . .* Oxoniae: Exudebant I. Lichfield & I. Short, 1621.

van Helmont, Jean Baptist. *Oriatrike, or Physick refined.* Translated by John Chandler. London: Lodowick Loyd, 1662.

Webster, John. *The Displaying of Supposed Witchcraft . . .* London: Jonas Moore, 1677.

Zetzner, Lazar. *Theatrum Chemicum.* 2nd ed. 5 vols. Argentorati [Strasbourg]: Sumptibus Heredum Eberh. Zetzneri, 1659.

D. Documentary Collections

British Record Society. *The Index Library.* London: British Record Society, 1888– .

Great Britain. Historical Manuscripts Commission. *Reports.* Series 1–81. London: H. M. Stationery Office, 1870–1946.

——. Public Record Office. *Calendar of Patent Rolls, Elizabeth I (1558–1572).* 5 vols. London: H. M. Stationery Office, 1939–1966.

——. ——. *Calendar of State Papers, Domestic Series, of the Reigns of Edward VI, Mary, Elizabeth and James I.* 12 vols. London: H. M. Stationery Office, 1856–1872.

——. ——. *Calendar of State Papers, Domestic Series, of the Reign of Charles I.* 23 vols. London: H. M. Stationery Office, 1858–1897.

——. ——. *Calendar of State Papers, Foreign Series, of the Reign of Elizabeth.* 22 vols. London: H. M. Stationery Office, 1863–1950.

——. ——. *Privy Council Registers. Acts of the Privy Council of England.* 42 vols. (1542–1627). London: H. M. Stationery Office, 1890–1938.

Harleian Society, London. *Publications.* 91 vols. London: The Harleian Society, 1869–1939.

IV. SECONDARY SOURCES

A. Books on Fludd

Craven, James Brown. *Doctor Robert Fludd.* Kirkwall: William Peace and Son, 1902.

Godwin, Joscelyn. *Robert Fludd: Hermetic Philosopher and Surveyor of Two Worlds.* Boulder, Colorado: Shambhala and London: Thames and Hudson, 1979.

Hutin, Serge. *Robert Fludd (1574–1637).* Paris: Omnium Littéraire, 1971.

B. Other Books and Articles

Allen, D. C. *Mysteriously Meant.* Baltimore and London: Johns Hopkins Press, 1970.

Allen, Paul M., ed. *A Christian Rosenkreutz Anthology.* Blauvelt, N.Y.: Rudolf Steiner, 1968.

Ammann, Peter J. "The Musical Theory and Philosophy of Robert Fludd," *Journal of the Warburg and Courtauld Institutes,* 30 (1967), 198–227.

"An Historical Account of the Origin and Establishment of the Society of Antiquaries," *Archaeologia,* 1 (1770). i–xxxix.

Barton, Todd. "Robert Fludd's *Temple of Music*: A Description and Commentary." Unpublished M.A. Thesis, University of Oregon, 1978.

Bayon, H. P. "William Gilbert, Robert Fludd and William Harvey as Medical Exponents of Baconian Doctrines," *Proceedings of the Royal Society of Medicine,* 32 (1938–9), 31–42.

Bernheimer, Richard. "Another Globe Theatre," *Shakespeare Quarterly,* 9 (1958), 19–29.

Berry, Herbert. "Dr. Fludd's Engravings and Their Beholders," *Shakespeare Studies,* 3 (1967), 11–21.

Blau, Joseph L. *The Christian Interpretation of the Cabala in the Renaissance.* Port Washington, N.Y.: Kennikat Press, 1965.

Bonelli, M. L. R. and Shea, W. R., eds. *Reason, Experiment and Mysticism in the Scientific Revolution*. New York: Science History Publications, 1975.

Bowen, C. D. *Francis Bacon*. Boston and Toronto: Little, Brown, 1963.

Bouwsma, William J. *Concordia Mundi: The Career and Thought of Guillaume Postel (1510–1581)*. Cambridge: Harvard University Press, 1957.

Burkhardt, Titus. *Alchemy*. Baltimore: Penguin Books, 1967.

Burtt, E. A. *The Metaphysical Foundations of Modern Science*. Garden City, N.Y.: Doubleday, 1954.

Calder, I. R. F. "John Dee Studied as an English Neo-Platonist." Thesis, University of London, 1952.

Caspari, Fritz. *Humanism and the Social Order in Tudor England*. New York: Teacher's College Press, 1968.

Cafiero, L. "Robert Fludd e la polemica con Gassendi," *Revista Critica di Storia Filosofia* 19 (1964), 367–410; 20 (1965), 3–15.

Cassirer, Ernst. *The Individual and the Cosmos in Renaissance Philosophy*. Translated with an introduction by Mario Domandi. Philadelphia: University of Pennsylvania Press, 1972.

———. *The Platonic Renaissance in England*. New Haven: Yale University Press, 1953.

———. *et al*, eds. *The Renaissance Philosophy of Man*. Chicago: University of Chicago Press, 1948.

Church, R. W. *Bacon*. New York: AMS Press, 1968.

Clark, Andrew, ed. *Register of the University of Oxford*. 4 vols. Oxford Historical Society Publications X–XIV. Oxford: Clarendon Press, 1887–1889.

Clark, George. *A History of the Royal College of Physicians of London*. 2 vols. Oxford: Clarendon Press, 1964.

Collier, K. B. *Cosmologies of Our Fathers*. New York: Columbia University Press, 1934.

Conger, G. P. *Theories of Macrocosms and Microcosms*. New York: Columbia University Press, 1922.

Costin, W. C. *The History of St. John's College, Oxford, 1598–1860*. Oxford Historical Society, New Series, XII. Oxford: Clarendon Press, 1958.

———. "The Inventory of John English, B.C.L., Fellow of St. John's College," *Oxoniensia*, 11–12 (1946–7), 102–31.

Craven, James Brown. *Count Michael Maier*. Kirkwall: William Peace and Son, 1910.

Curtis, Mark H. *Oxford and Cambridge in Transition, 1558–1642*. Oxford: Clarendon Press, 1959.

Davies, D. W. *Elizabethans Errant*. Ithaca: Cornell University Press, 1967.

Deacon, Richard. *John Dee*. London: Frederick Muller, 1968.

Debus, Allen G. "The Chemical Debates of the Seventeenth Century," *Reason, Experiment and Mysticism in the Scientific Revolution*. New York: Science History Publications, 1975.

———. *The Chemical Dream of the Renaissance*. Cambridge, England: W. Heffer and Sons, 1968.

———. *The Chemical Philosophy: Paracelsian Science and Medicine in the Sixteenth and Seventeenth Centuries*. New York: Science History Publications, 1977.

———. *The English Paracelsians*. New York: Franklin Watts, 1965.

——. "Harvey and Fludd: The Irrational Factor in the Rational Science of the Seventeenth Century," *Journal of the History of Biology*, 3 (1970), 81–105.

——. *Man and Nature in the Renaissance*. Cambridge: Cambridge University Press, 1978.

——. "The Paracelsian Aerial Niter," *Isis*, 55 (1964), 43–61.

——. "Renaissance Chemistry and the Work of Robert Fludd," *Ambix*, 14 (1967), 42–59.

——. "Robert Fludd and the Circulation of the Blood," *Journal of the History of Medicine*, 16 (1961), 374–93.

——. "Robert Fludd and the Use of Gilbert's *De Magnete* in the Weapon-Salve Controversy," *Journal of the History of Medicine and Allied Sciences*, 19 (1964), 389–417.

——. *Science and Education in the Seventeenth Century: The Webster-Ward Debate*. New York: Science History Publications, 1970.

——. "The Sun in the Universe of Robert Fludd," *Le Soleil à la Renaissance*. Brussels: Presses Universitaires de Bruxelles, 1965.

Dictionary of National Biography. s. v. "Fludd, Robert," by Alexander Gordon.

Dictionary of Scientific Biography. s. v. "Fludd, Robert," by Allen G. Debus.

D'Israeli, I. *Amenities of Literature*. 4th ed. 2 vols. New York: Harper, 1855.

Dobbs, B. J. T. *The Foundations of Newton's Alchemy*. Cambridge: Cambridge University Press, 1975.

Durling, Richard J. "Some Unrecorded Verses in Praise of Robert Fludd and William Harvey," *Medical History* 8 (1964), 279–81.

Eleade, Mircea. *The Forge and the Crucible: Origins and Structures of Alchemy*. New York: Harper and Row, 1971.

Encyclopedia of Philosophy. s. v. "Fludd, Robert," by John Passmore.

Evans, Joan. *A History of the Society of Antiquaries*. Oxford: The University Press, 1956.

Evans, R. J. W. *Rudolf II and His World*. Oxford: Clarendon Press, 1973.

Festugière, André. *La révélation d'Hermès Trismégiste*. 4 vols. Paris: J. Gabalda, 1949–1954.

Ferguson, John. *Bibliotheca Chemica*. 2 vols. Glasgow: James Maclehose, 1906. Reprint, London: Derke Verschoyle, 1954.

Field, J. V. *Kepler's Geometrical Cosmology*. Chicago: University of Chicago Press, 1988.

Forneron, H. *Les Ducs de Guise et leur époque*. 4 vols. Paris: E. Plon, 1877.

Foster Joseph, ed. *The Register of Admissions to Gray's Inn, 1521–1889*. London: Privately Printed, 1889.

French, Peter J. *John Dee: The World of an Elizabethan Magus*. London: Routledge and Kegan Paul, 1972.

Freudenberg, Fr. *Paraculsus und Fludd*. Berlin: Hermann Barsdorf, 1918.

Fuller, Thomas. *The History of the Worthies of England*. 3 vols. Privately Printed, 1923.

Godwin, Joscelyn. "Instruments in Robert Fludd's *Utriusque Cosmi . . . Historia*," *Galpin Society Journal*, 26 (1973),

——. "Robert Fludd on the Lute and Pandora," *Lute Society Journal*, 15 (1973), 11–19.

——. *Robert Fludd: Hermetic Philosopher and Surveyor of Two Worlds* London: Thames and Hudson, 1979.

Gough, J. W. *The Rise of the Entrepreneur*. New York: Schocken Books, 1969.

Granger, James, ed. *A Biographical History of England*. 2nd ed. 4 vols. London: T. Davis, 1775.

Gunther, T. T. *Early Science in Oxford*. 2 vols. Oxford Historical Society 77–8. Oxford: Clarendon Press, 1923.

Guthrie, Kenneth S., Compiler and Translator. *The Pythagorean Sourcebook and Library*. Grand Rapids, Michigan: Phanes Press, 1987.

Hall, Manly P. *Codex Rosae Crucis*. Los Angeles: Philosophical Research Society, 1974.

——. *Man, the Grand Symbol of the Mysteries*. Los Angeles: Manley Hall Publications, 1932.

Haller, Albrecht von. *Bibliotheca Medicinae Practicae*. Bern: E. Haller, 1777.

Hasted, Edward. *History of Topographical Survey of the County of Kent*. 4 vols. Canterbury: Simmons and Kirby, 1778–1790.

Henninger, S. K. *The Cosmographical Glass*. San Marino: Huntington Library, 1977.

Hermes Trismegistus. *Corpus Hermeticum*. Edited by A. D. Nock, translated by A. J. Festigière. 4 vols. Paris: Société d'Édition "Les Belles Lettres," 1945–1954.

——. *Hermetica*. Edited and translated by Walter Scott. 4 vols. Oxford: Clarendon Press, 1924–1936.

——. *The Divine Pimander*. Translated by John D. Chambers. New York: Weiser, 1972.

Hirst, Desirée. *Hidden Riches: Traditional Symbolism From the Renaissance to Blake*. London: Eyre and Spottiswoode, 1964.

Holmyard, E. J. *Alchemy*. Harmondsworth, England: Penguin, 1968.

Howell, Wilbur S. *Logic and Rhetoric in England, 1500–1700*. Princeton: Princeton University Press, 1956.

Hutton, William H. *S. John Baptist College*. London: F. E. Robinson, 1898.

Janson, H. W. *Apes and Ape Lore in the Middle Ages and the Renaissance*. London: The Warburg Institute, 1952.

Jayne, Sears. *Library Catalogues of the English Renaissance*. Berkeley and Los Angeles: University of California Press, 1956.

Johnson, F. R. *Astronomical Thought in Renaissance England*. New York: Octagon, 1968.

Josten, C. H. "A Translation of John Dee's 'Monas Hieroglyphica,'" *Ambix*, 12 (1964), 84–221.

——. ed. *Elias Ashmole*. 5 vols. Oxford: The Clarendon Press, 1966.

——. "Robert Fludd's 'Philosophical Key' and his Alchemical Experiment of Wheat," *Ambix*, 11 (1963), 1–26.

——. "Robert Fludd's Theory of Geomancy and his Experiences at Avignon in the Winter of 1601 to 1602," *Journal of the Warburg and Courtauld Institutes*, 27 (1964), 327–335.

——. "Truth's Golden Harrow: An Unpublished Alchemical Treatise of Robert Fludd," *Ambix*, 3 (1949), 91–150.

Kearney, Hugh. *Science and Change 1500–1700.* New York: World University Library, 1971.

Keynes, Geoffrey. *The Life of William Harvey.* Oxford: The Clarendon Press, 1966.

Kinney, A. F. *Titled Elizabethans.* Hamden, Conn.: Archon, 1973.

Kohler, Richard C. "The Fortune Contract and Vitruvian Symmetry," *Shakespeare Studies*, 6 (1966), 192–209.

Koestler, Arthur. *The Watershed: A Biography of Johannes Kepler.* Lanham, Maryland: University Press of America, 1985.

Kristeller, Paul Oskar. *The Philosophy of Marsilio Ficino.* Translated by Virginia Conant. New York: Columbia University Press, 1943. Reprint, Gloucester, Mass: Peter Smith, 1964.

——. *Renaissance Thought and Its Sources.* New York: Columbia University Press, 1979.

——. "Giovanni Pico della Mirandola and His Sources." *L'Opera e il Pensiero di Giovanni Pico della Mirandola.* 2 vols. Florence: Istituto Nazionale di Studi sul Rinascimento, 1965.

Kuhn, Thomas S. *The Structure of Scientific Revolutions.* 2nd Edition. Chicago: University of Chicago Press, 1970.

Lenoble, Robert. *Mersenne ou la Naissance du Méchanisme.* Paris: J. Vrin, 1943.

Lovejoy, Arthur O. *The Great Chain of Being.* Cambridge: Harvard University Press, 1964.

Mallet, Charles E. *A History of the University of Oxford.* 3 vols. New York: Longmans, Green, 1924–1927.

Manget, Jean Jacques. *Bibliotheca scriptorum medicorum . . .* 2 vols. Geneva: Achon & Cramer, 1731.

McLean, Antonia. *Humanism and the Rise of Science in Tudor England.* New York: Neale Watson, 1972.

Montgomery, John Warwick. *Cross and Crucible: Johann Valentin Andreae (1586–1654).* 2 vols. The Hague: Martinus Nijhoff, 1973.

Munk, William. *The Roll of the Royal College of Physicians of London.* 2 vols. London: Longmans, Green, 1861.

Nauert, Charles G., Jr. *Agrippa and the Crisis of Renaissance Thought.* Urbana: University of Illinois Press, 1965.

Pady, Donald S. "Sir William Paddy, M.D. (1554–1634)," *Medical History*, 18 (1974), 68–82.

Pagel, Walter. *Das medizinische Weltbild des Paracelsus: seine Zusammenhange mit Neuplatonismus und Gnosis.* Wiesbaden: Franz Steiner Verlag, 1962.

——. *New Light on William Harvey.* Basel and New York: S. Karger, 1976.

——. "Paracelsus and the Neoplatonic and Gnostic Tradition," *Ambix*, 8 (1960), 125–66.

——. *Paracelsus: An Introduction to Philosophical Medicine in the Era of the Renaissance.* Basel and New York: S. Karger, 1958.

——. "Religious Motives in the Medical Biology of the Seventeenth

244

Century," *Bulletin of the Institute of the History of the Medicine*, 3 (1935), 97–312.

———. *William Harvey's Biological Ideas*. Basel and New York: S. Karger, 1967.

Paracelsus, Theophrast von Hohenheim. *Selected Writings*. Edited with an introduction by Jolande Jacobi. Translated by Norbert Guterman. Bollingen Series 28. Princeton: Princeton University Press, 1958.

———. *Sammtliche Werke. 1. Abteilung: Medizinische, naturwissenschaftliche und philosophische Schriften*. Edited by Karl Sudhoff. 14 vols. Munich: R. Oldenbourg, 1922–1933.

———. *The Hermetic and Alchemical Writings of Paracelsus*. Edited by Arthur Edward Waite. 2 vols. London: James Elliot, 1894. Reprint, Berkeley: Shambhala, 1976.

Pauli, Wolfgang. "The Influence of Archetypal Ideas on the Scientific Theories of Kepler," *The Interpretation of Nature and the Psyche*. Bollingen Series 51. New York: Pantheon Books, 1955.

Peuckert, Will-Erich. *Die Rosenkreutzer*. Jena: Eugen Diederichs, 1928.

———. *Pansophie*. Berlin: Erich Schmidt, 1956.

Popkin, Richard H. *The History of Scepticism from Erasmus to Descartes*. New York: Humanities Press, 1960.

Purnell, Frederick, Jr. "Francesco Patrizi and the Critics of Hermes Trismegistus," *The Journal of Medieval and Renaissance Studies*, 6 (1976), 155–78.

Rattansi, P. M. "Paracelsus and the Puritan Revolution," *Ambix*, 11 (1963), 24–32.

Read, John. *Prelude to Chemistry: An Outline of Alchemy*. Cambridge, Mass.: Massachusetts Institute of Technology Press, 1966.

Rees, Graham. "The Fate of Bacon's Cosmology in the Seventeenth Century," *Ambix*, 24 (1977), 27–38.

Rhodes, E. L. "Cleopatra's 'Monument' and the Gallery in Fludd's *Theatrum Orbi*," *Renaissance Papers*. n. p.: Southeast Renaissance Conference, 1972.

Robb, Nesca A. *Neoplatonism of the Italian Renaissance*. New York: Octagon Books, 1968.

Roberts, R. S. "The Personnel and Practice of Medicine in Tudor and Stuart England. Part II, London," *Medical History*, 8 (1964), 217–34.

Rommel, Christoph von. *Geschichte von Hessen*. 9 vols. Cassel: Hampe, *et al*, 1820–1853.

Rossi, Paolo. *Francis Bacon: From Magic to Science*. Chicago: University of Chicago Press, 1968.

———. *Philosophy, Technology and the Arts in the Early Modern Era*. Harper and Row, 1970.

Røstvig, Maren-Sofie. "'The Rime of the Ancient Mariner' and the Cosmic System of Robert Fludd," *Tennessee Studies in Literature*, 12 (1967), 69–82.

Rowse, A. L. *Simon Forman: Sex and Society in Shakespeare's Age*. New York: Scribner's, 1974.

Saurat, Denis. *Literature and Occult Tradition*. London: Bell, 1930.

——. *Milton et le Matérialisme Chrétien en Angleterre*. Paris: Rieder, 1928.

——. *Milton, Man and Thinker*. New York: Dial, 1925.

Schneider, Heinrich. *Joachim Morsius und sein Kreis: zur Geistesgeschichte des 17. Jahrhunderts*. Lübeck: Otto-Quitzow, 1929.

Schlenz, Hermann. "Goldmachen und Goldmacher am hessischen Hofe," *Deutsche Geschichtsblätter*, 11 (1910), 308–11.

Scholem, Gershom G. *Major Trends in Jewish Mysticism*. New York: Schocken Books, 1961.

——. *On the Kabbalah and its Symbolism*. Translated by Ralph Mannheim. New York: Schocken Books, 1969.

Secret, F. *Les Kabbalistes Chrétiens de la Renaissance*. Paris: Dunod, 1964.

Seligmann, Kurt. *Magic, Supernaturalism and Religion*. New York: Pantheon Books, 1971.

Shapiro, I. A. "Robert Fludd's Stage-Illustration," *Shakespeare Studies*, 2 (1966), 192–209.

Shaw, William A. *The Knights of England*. 2 Vols. London: Sherratt and Hughes, 1906.

Shumaker, Wayne. *The Occult Sciences in the Renaissance*. Berkeley and Los Angeles: University of California Press, 1972.

Shrewsbury, J. F. D. *A History of Bubonic Plague in the British Isles*. Cambridge: The University Press, 1970.

Simmonds, Mark J. ed. *Merchant Taylor Fellows of St. John's College, Oxford*. London: Humphrey Milford, 1930.

Singleton, C. S. ed. *Art, Science and History in the Renaissance*. Baltimore and London: Johns Hopkins Press, 1967.

Smith, Charlotte Fell. *John Dee (1527–1608)*. London: Constable, 1909.

Stephenson, W. H. and Walter, H. E., *The Early History of St. John's College, Oxford*. Oxford Historical Society, New Series, no. 1. Oxford: Clarendon Press, 1939.

Taylor, F. Sherwood. "The Origin of the Thermometer," *Annals of Science* 5 (1942), 129–156.

Teich, Mikulas and Young, Robert, eds. *Changing Perspectives in the History of Science*. London: Heinemann, 1973.

Temple, William. *Sir William Temple's Essays On Ancient and Modern Learning and On Poetry*. Edited by J. E. Springarn. Oxford: Clarendon Press, 1909.

Thomas, Keith. *Religion and the Decline of Magic*. New York: Scribner's, 1971.

Thompson, William Irwin. *At the Edge of History*. New York: Harper and Row, 1971.

——. *Darkness and Scattered Light*. Garden City, N.Y.: Anchor/Doubleday, 1978.

Thorndike, Lynn. *A History of Magic and Experimental Science*. 8 vols. New York: Columbia University Press, 1923–1958.

Tillyard, E. M. W. *The Elizabethan World Picture*. New York: Vintage, n.d.

Trevor-Roper, H. R. *Archbishop Laud, 1573–1645*. London: Macmillan, 1940.

——. *The Gentry 1540–1640*. London: Cambridge University Press, n.d.

Vickers, Brian, ed. *Occult and Scientific Mentalities in the Renaissance*. Cambridge: Cambridge University Press, 1984.

Waite, Arthur Edward. *The Alchemical Writings of Edward Kelley.* New York: Weiser, 1973.

——. *The Brotherhood of the Rosy Cross.* New Hyde Park, N.Y.: University Books, 1961.

——. "Haunts of the English Mystics," *Square and Compass* 47 (1933), 11–15.

——. *Hermetic Museum.* 2 vols. New York: Weiser, 1974.

——. *The Holy Kabbalah.* New Hyde Park, N.Y.: University Books, 1960.

——. *The Real History of the Rosicrucians.* London: Redway, 1887.

——. "Robert Fludd: Philosopher and Occultist," *The Occult Review* 15 (1912), 79–84.

——. *The Works of Thomas Vaughn.* New Hyde Park, N.Y.: University Books, 1968.

Wall, Cecil, *et al. A History of the Worshipful Society of Apothecaries of London.* 2 vols. London: Oxford University Press, 1963.

Wallis, R. T. *Neoplatonism.* New York: Scribner's, 1972.

Walker, D. P. *The Ancient Theology.* Ithaca: Cornell University Press, 1972.

——. "The Astral Body in Renaissance Medicine," *Journal of the Warburg and Courtauld Institutes,* 21 (1958), 119–33.

——. *Spiritual and Demonic Magic from Ficino to Campanella.* London: The Warburg Institute, 1958.

Ward, John. *The Lives of the Professors of Gresham College.* New York: Johnson Reprint, 1967.

Westfall, R. S. *Science and Religion in Seventeenth Century England.* Ann Arbor: University of Michigan Press, 1973.

Westman, Robert S. and McGuire, J. E. *Hermeticism and the Scientific Revolution.* Los Angeles: William Andrews Clark Library, 1977.

Whitterridge, Gwenneth. *William Harvey and the Circulation of the Blood.* New York: American Elsevier, 1971.

Wightman, W. P. D. *Science in a Renaissance Society.* London: Hutchinson University Library, 1972.

Wilkins, John. *The Mathematical and Philosophical Works of the Right Reverend John Wilkins.* London: Class, 1970.

Willey, Basil. *The Seventeenth Century Background.* Garden City, N.Y.: Doubleday, 1953.

Willson, D. Harris. *King James VI and I.* New York: Henry Holt, 1956.

Wind, Edgar. *Pagan Mysteries in the Renaissance.* 2nd ed. New York: Norton, 1968.

Wolstenholm, Gordon, ed. *The Royal College of Physicians: Portraits.* London: Churchill, 1964.

Wood, Anthony. *Athenae Oxonienses,* ed. Philip Bliss. 4 vols. London: J. Rivington, *et al,* 1813–1820.

Yates, Frances A. *The Art of Memory.* Chicago: University of Chicago Press, 1966.

——. *Giordano Bruno and the Hermetic Tradition.* Chicago: University of Chicago Press, 1964.

——. "The Hermetic Tradition in Renaissance Science," *Art, Science and History in the Renaissance.* Edited by Charles Singleton. Baltimore and London: Johns Hopkins Press, 1967.

——. *The Occult Philosophy in the Elizabethan Age*. London and Boston: Routledge and Kegan Paul, 1979.

——. *The Rosicrucian Enlightenment*. London and Boston: Routledge and Kegan Paul, 1972.

——. "The Stage in Robert Fludd's Memory System," *Shakespeare Studies* 3 (1967), 138–66.

——. *Shakespeare's Last Plays*. London: Routledge and Kegan Paul, 1975.

——. *Theater of the World*. Chicago: University of Chicago Press, 1969.

INDEX

INDEX

Promethius, 120
Psellus, 91
Ptolemy, Cladius, 77
Pythagoras, 40, 53, 54, 55, 60, 75, 77, 89, 90, 91, 92, 96, 100, 102, 103, 140, 155, 171, 212

Ramus, Petrus, 12
Reginaud of Avignon, 30
Reuchlin, Johann, 21, 33, 97, 128
Ridley, Mark, 100, 125, 175
Rochier, John, 24, 46, 47
Rosicrucians, 2, 20, 36ff., 56, 71, 77, 135ff., 169, 174, 176, 177, 178, 206, 207, 210ff.
Rudolf II, Holy Roman Emperor, 31, 54, 153, 164, 170

Saccus, Ammonius, 79
Seelinger, Robert A., 209
St George, Cardinal, 30
Selden, John, 33, 34, 151, 205, 206
Sennert, Daniel, 24
Scott, Patrick, 62
Scotus, Duns, 86, 91
Shakespeare, 105, 134
Sibylline Prophecies, 89
Silkeman, Kenneth 227
Society of Antiquaries, 33
Socrates, 74, 92, 103
Solomon, 140
Sumum Bonum, 63, 64, 157, 158, 159, 160, 207
Sophie cum Moria Certamen, 63, 64, 157, 207
Suso, Blessed Henry, 86
Studion, Simon, 162

Tauler, John, 86
Temple, Sir William, 175
Thales of Melissus, 40, 212
Thompson, William Irwin, 168

Thorborough, Sir Thomas, 32, 207
Thornborough, John, Bishop of Worcester, 11, 32, 50, 207
Tractatus Apologeticus, 27, 37, 39, 41, 51, 145, 206, 209ff.
Tractatus Theologo-Philosophicus 32, 146, 150, 153, 206
Trevor, Arthur, Esq., 223, 227
Trinity College, Cambridge, 45
Trithemius, 63, 159
Truth's Golden Harrow, 62
Typhon, 120

Utriusque cosmi . . . historia, 29-30, 34, 36, 37, 41, 42, 43, 44, 51, 52, 55, 69, 99, 136, 148, 155, 205, 206, 207, 209, 214ff., 222, 232

van Helmont, 20
van Ruysbroeck, Jan, 86
Vaughn, Katherine, 226
Veritatis Proscenium, 52, 207
Victorinus, Marius, 83

Waite, A. E., 178-9
Ward, Katherine, 226
Weapon salve, 22
Weather-Glass, 121ff.
Wessel, William, 164
Westman, Robert S., 59ff., 161
William IV the Wise, Landgrave of Hesse-Cassel, 31, 164
Williams, John, Bishop of Lincoln, Archbishop of York, 11, 50, 207
Wood, Anthony, 171, 176-7, 178
Wotten, Sir Henry, 62
Wright, Factor junior, 20

Yates, Dame Frances A., 12, 65, 89, 134, 135, 156, 160, 161, 162, 179

Zoraster, 89, 90, 119

DUE DATE

	201-6503		Printed in USA